I REALIZED GOD THROUGH MATHEMATICS

Serge Raynaud de la Ferrière: His Primordial Thought

by

David Ferriz

The first book of a series presenting
the primordial thought
of
Serge Raynaud de la Ferrière

Translated by David Ferriz
Assistance in translation was provided
by Joseph Patrick, Polyglot Instructor
Title page by Roger Harida & Sara Peche
Coordination and direction in this edition
of the work: Dr. David Ferriz
Assistance in coordination and direction
of the work was provided by
Graciela Fusaro and F. Salomon.
Appreciation is also extended to UGB members
in Huntington Park, California,
New York City, and Miami, Florida
for their economic and moral support.

Distribution Centers:

Universal Great Brotherhood
P.O. Box 516
Huntington Park, CA 90255

Universal Great Brotherhood
P.O. Box 949
New York, NY 10001

Hwong Publishing Company
10353 Los Alamitos Blvd.
Los Alamitos, CA 90720
U.S.A.

I Realized God Through Mathematics

A Translation of the revised Spanish edition.

ISBN 0-89260-191-4

DOCUMENTATION

This work was presented to the Director General of UNESCO in Paris for their collection by the Vice President of the National Council of the Peruvian University, an organization representing thirty-three universities. The work was also sponsored for publication by the National University of Trujillo, Peru for the University bibliography of the organizations of Education, Science and Culture.

* * *

The following letter shows that the work,

I REALIZED GOD THROUGH MATHEMATICS
Serge Raynaud de la Ferrière: His Primordial Thought

was sponsored for publication for the University Bibliography of the Organizations of Education, Science and Culture by Rectoral Resolution No. 1653 from the National University of Trujillo in Peru.

UNIVERSIDAD NACIONAL DE TRUJILLO

RECTORADO

INDEPENDENCIA 431 - OF. 203.

Trujillo, 19 de Noviembre de 1976

RESOLUCION RECTORAL N° 1653.-

CONSIDERANDO:

Que, merecida la más alta atención a la obra "SERGE RAYNAUD de la FERRIERE, SU PENSAMIENTO PRIMORDIAL", del investigador y epistemólogo David Ferriz Olivares, quien se ubica en un alto nivel del pensamiento, estableciendo un arco de unión a escala mundial, al extender, con base en una cuantiosa documentación, su saber de Voz Universal ;

Que, los planteamientos de investigación y de vivencia aplicados a la realización de la CAUSA SUPREMA a través de las matemáticas, al descubrimiento del Hombre Trascendental, a la transición entre dos Eras, así como a otros temas de previsible evolución de avanzada pueden ser emprendidos y ensanchados por medio de esta serie de libros y colectivamente ;

Que el binomio que establece entre los eminentes Serge Raynaud de la Ferriere y Julián Huxley, fundadores de la GFU y de la UNESCO respectivamente, abre perspectivas de síntesis entre las instituciones de autorización Ontológica de Educación, de Ciencia y de Cultura ;

Que el seguimiento exacto del pensamiento de Serge Raynaud de la Ferriere, como el mismo lo declaró para su traductor e ideólogo David Ferriz Olivares proporciona una validéz especial a la investigación y explicación del pensamiento esencial del sabio francés, supera los niveles simplistas con criterios de Cosmovisión Académica y Científica, y hará que el eminente sabio francés sea más conocido en Europa ;

Que dentro del plan de la obra en mención se abren objetivos de aplicación curricular ;

UNIVERSIDAD NACIONAL DE TRUJILLO

RECTORADO

INDEPENDENCIA 431 - OF. 203

S E R E S U E L V E :

ARTICULO UNICO : AUSPICIAR la publicación de la obra "SERGE RAYNAUD de la FERRIERE, SU PENSAMIENTO PRIMORDIAL" del doctor David Ferriz Olivares y recomendar su consideración en la bibliografía universitaria de los organismos de la Educación, la Ciencia y la Cultura.

Regístrese, comuníquese y archívese.

ANIBAL ESPINO RODRIGUEZ
RECTOR

CARLOS M. CENTURION CUEVA
PROFESOR-SECRETARIO GENERAL

UNIVERSIDAD NACIONAL DE TRUJILLO
Secretaría General

JAC.Lgc.

NATIONAL UNIVERSITY OF TRUJILLO

DIRECTORATE

INDEPENDENCIA 431 - OF. 203.

Trujillo, November 19, 1976

RECTORAL RESOLUTION No. 1653.

IT OUGHT TO BE CONSIDERED:

That, the highest attention is merited by the work, *SERGE RAYNAUD de la FERRIERE—HIS PRIMORDIAL THOUGHT*, by the researcher and epistemologist, David Ferriz Olivares, who is situated on the highest level of thought, establishing an arch of union to the world level, and extending his knowledge of the Universal Voice based on plentiful documentation;

That, the outlines of investigation and of living experience applied to the realization of the SUPREME CAUSE through Mathematics, to the discovery of Transcendental Man, to the transition among two Ages, likewise to other themes of previsible development of a vanguard able to be undertaken and extended by means of this series of books and, collectively;

That the binomial which it establishes between the eminent Serge Raynaud de la Ferrière and Julian Huxley, founders of the UGB and UNESCO, respectively, opens up perspectives of synthesis between the institutions of authorization ontologically of Education, Science, and Culture;

That the exact following of the thought of Serge Raynaud de la Ferrière as he declares it himself for his translator and ideologue, David Ferriz Olivares adapts a special validity to the investigation and explanation of the essential thought of the French sage, surpassing the simplistic levels—with criteria of Academic and Scientific Cosmovision, and will make the eminent French Sage more known in Europe;

That within the plan of the work in reference to objectives up of curricular applications;

BE IT RESOLVED:

ARTICLE ONE: TO SPONSOR the publication of the work, *SERGE RAYNAUD de la FERRIÈRE, HIS PRIMORDIAL THOUGHT* of Doctor David Ferriz Olivares and to recommend its consideration in the Universitary Bibliography of the organizations of Education, Science, and Culture.

To be registered, communicated, and filed.

ANIBAL ESPIÑO RODRIGUEZ
RECTOR

CARLOS M. CENTURION CUEVA
PROFESSOR-SECRETARY GENERAL

JAC.Lgc.

The following letter shows that the work was presented to the Director General of UNESCO in Paris for their collection by the Vice-President of the National Council of the Peruvian University.

"AÑO DE LA PRODUCCION"

CONSEJO NACIONAL DE LA UNIVERSIDAD PERUANA

Calle ALDABAS – Cdra. 3
LIMA - 33 - PERU
Telf. 56–34–91
Apartado 4664
Cables: CONUP

Lima, 21 de Octubre de 1976.

Señor Doctor
Mahtar M'Bow
Director General de la UNESCO
Place Fontenoy 75700
PARIS- FRANCIA

De mi más alta consideración:

Desde la Vice-presidencia del Consejo Nacional de la Universidad Peruana, organismo representativo de 33 Universidades envío a Ud. nuestra cordial salutación, con motivo del XXX Aniversario de la creación de la UNESCO.

Nos es muy honroso presentar a su digna consideración para su publicación un libro que trata del pensamiento de uno de los mayores admiradores de la UNESCO, el ilustre francés Serge Raynaud de la Ferriére (1916-1962) ampliamente conocido en todo el Continente Americano.

Fue escrito por David Ferriz O, docto investigador y escritor de ciudadanía continental, nacido en Japón en 1921, hijo de padres mexicanos en servicio diplomático en aquel país, considerado por muchos jóvenes como Epistemólogo y Maestro de la Cultura Universal.

Agradeciendo anticipadamente su atención, le presento las seguridades de mi elevada consideración y estima personal.

Dr. Anibal Espino Rodriguez
VICE-PRESIDENTE DEL SISTEMA NACIONAL DE LA UNIVERSIDAD PERUANA.
RECTOR DE LA UNIVERSIDAD NACIONAL DE TRUJILLO P E R U.

"YEAR OF PRODUCTION"

NATIONAL COUNCIL OF THE PERUVIAN UNIVERSITY

Lima, October 11, 1976.

Doctor Mahtar M'Bow
Director General of UNESCO
Place Fontenoy 75700

PARIS, FRANCE

Of my highest consideration:

From the Vice-Presidency of the National Council of the Peruvian University, an organization representing 33 Universities, we send to you our cordial greeting, on the XXX Anniversary of the creation of UNESCO.

We are very honored to present to your worthy consideration for its publication a book which deals with the thought of one of the greatest admirers of UNESCO, the illustrious Frenchman, Serge Raynaud de la Ferrière, (1916-1962) widely known in all the American continent.

It was written by David Ferriz Olivares, doctor, researcher, and writer of continental citizenship, born in Japan in 1921, son of Mexican parents in diplomatic service in that country, considered by many young people as an Epistemologist and Master of Universal Culture.

Thanking you in advance for your attention, I present you the assurity of my highest consideration and personal esteem.

Dr. Anibal Espino Rodriguez
VICE-PRESIDENT OF THE NATIONAL SYSTEM OF THE PERUVIAN UNIVERSITY.

DIRECTOR OF THE NATIONAL UNIVERSITY OF TRUJILLO

P E R U.

The following letter acknowledges receipt by the Peruvian National Commission on Cooperation with UNESCO of the work: *Serge Raynaud de la Ferrière—His Primordial Thought*.

"AÑO DE LA UNION NACIONAL"

Comisión Nacional Peruana
de Cooperación con la "Unesco"
Apartado 4681
Lima - Perú

Lima, a 04 de noviembre de 1977

Oficio N°231 -SP/77

Señor Director General
de la Gran Fraternidad Universal
Doctor DAVID FERRIZ OLIVARES
CARACAS VENEZUELA

Por la presente, me es sumamente grato acusar recibo de la considerable obra SERGE RAINAUD DE LA FERRIERE SU PENSAMIENTO PRIMORDIAL, Obra que se constituye en un factor de comprensión internacional para ampliar la base popular de la comunicación en grupos culturales, en ellas se exaltan las figuras de dos Grandes Pensadores que la Humanidad ha tenido como son los Sabios JULIAN HUXLEY y SERGE RAINAUD DE LA FERRIERE, quienes son enaltecidos y dignificados Históricamente por sus aportes tan trascendentales para la humanidad en el campo de la Educación, la Ciencia y la Cultura.

Por lo que estamos muy complacidos que se haya considerado para formar parte de la Colección Bibliográfica de la UNESCO.

Agradeciendo la fina atención, aprovecho la oportunidad para reiterarle los sentimientos de mi mayor aprecio

Atentamente,

JUAN MERLO LENO
Secretario Permanente de la Comisión
Nacional Peruana de Cooperación con
LA UNESCO

SP/JML
lmhn

"YEAR OF THE NATIONAL UNION"

Peruvian National
Commission of
Cooperation with UNESCO

Lima, November 4, 1977

Official Communication No. 231-SP/77

Director General of the
Universal Great Brotherhood
Doctor David Ferriz Olivares
CARACAS VENEZUELA

By the present letter, I am highly pleased to acknowledge receipt of the considerable work, *SERGE RAYNAUD de la FERRIÈRE, HIS PRIMORDIAL THOUGHT*, a work which constitutes a factor in international comprehension for extending the popular base of communication among cultural groups. In this work the figures of the two Great Thinkers which Humanity has had, the Sages JULIAN HUXLEY and SERGE RAYNAUD de la FERRIÈRE, are exalted and dignified historically by their transcendental contributions for humanity in the field of Education, Science, and Culture.

Therefore, we are very pleased that it has been considered for forming part of the Bibliographical Collection of UNESCO.

Being grateful for the fine attention, I take this opportunity for reiterating to you the sentiments of my great appreciation.

Cordially,

JUAN MERLO LENO
Permanent Secretary of the
Peruvian National Commission
on Cooperation with UNESCO.

SP/JML
lmhn

DEDICATION

To the writers and thinkers, the scientists and humanists, the ideologues and exegetes, the researchers in general who broach and delve into the literature of transformation and learning (a literature also of investigation and of living experience) from the eminent Serge Raynaud de la Ferrière, imparting the union and interaction between his thought and the world thought, for the today and the tomorrow of this New Age.

David Ferriz

CONTENTS

INTRODUCTION

Concerning the analysis of the socio-educative aspect, it is necessary to sound the depths of those thoughts which gave rise to the founding of the Universal Great Brotherhood by Dr. Serge Raynaud de la Ferrière and to understand those levels to be reached by science education and culture, thus locating an exact point from which to observe that other great organization, UNESCO. For this reason a comparison is made between his thought and that of Julian Huxley.

Although these two figures were not personally acquainted, it is interesting to note the similarities in their objectives and the role of UNESCO in establishing the framework of this book's theme. Although UNESCO *as an institution* will survive for some time to come, it is the aims and the objectives that will serve as the mold and spur for such organizations as Dr. Raynaud de la Ferrière's Universal Great Brotherhood, and that will, across the years, cause a progressive evolution towards those objectives originated by the founder.

When will UNESCO reach its objectives? When will the UGB reach that level of education and culture planned for it, at the height occupied by UNESCO? Will the overly governmental and bureaucratic evolution consider the necessity of developing the high ends of UNESCO even on the popular level? Will the members of the UGB from their base on the popular level, take on the high level of organization which its founder had in mind for the institution? Will the plans of Dr. Serge Raynaud de la Ferrière and Julian Huxley be soon recognized by history and by the social, cultural and spiritual evolution of the peoples of the future?

It is interesting to note the points at which the beliefs of both coincide, especially as regards modern mysticism, sustained to some extent by scientific thought. In 1946 shortly before the official founding of UNESCO, Huxley wrote an important paper entitled "UNESCO: ITS OBJECTIVES AND PHILOSOPHY." The controversy which arose due to the supposed 'anti-religious' tendencies of the document were such that the Preparatory Commission and the first General Conference of the organization could not approve its publication.

What did they think about their great organizations? How did these two men found and bring into being two organizations which turned out to be historically coincidental?

Huxley, among other things, planned a Governmental UNESCO which would cause communities to flower; Dr. Serge Raynaud de la Ferrière planned the U.G.B. which, from individual and collective levels and using methods of human perfection through the knowledge of synthesis and mathesis, would develop at the higher levels of science, philosophy, art and didactic, and attain the union of sciences and religions, namely: a world united by knowledge.

UNESCO has the means, but at times the objectives become cloudy. The U.G.B. has the objectives, lacking at times, the means.

The renowned historian Paul E. de Berredo Carneiro says that in Huxley, the man of science, the philosopher, the apostle, all take their corresponding places among the leading personalities of our time. He adds that posterity will judge Huxley and his work with gratitude and admiration and will come to see him, as he himself said: "If I am to be remembered, I hope it will not be for my specialized scientific work, but as a 'Generalist' i.e., someone who, paraphrasing and expanding on the phrase of Terence, is a person to whom nothing is foreign that is human and that exists in nature."

In the words of Matila Ghyka: 'Harmonizing or filling the interval between two given terms consists in finding the means which gives birth to proportion.' Plato uses these terms indifferently be it in the realm of mathematics, music, or cosmogony. The problem of harmony as expressed in *The Republic* generally consists of putting the intervals into proportion by means of terms which are given in definite relation with the initial terms. Thus the accordance of the intervals is obtained, i.e. interjecting the middle term in a syllogism, stacking up a chain of syllogisms in sorites, and thus placing a bridge between two islands of consonance, as though relating two prosaic images by the precise metaphor, joining by proportion and harmony based on the analogy of surface forms and architectonic volumes. This is stated by Plato himself, in the Timaeus, and explained very clearly by Vitruvius: 'All these operations are parallel and are analogous to the creation of musical harmony which is the preferred model or example of the Pythagoreans'.

Naturally, it is necessary to consider the mathematical environment in both the ancient and modern social structure in order to direct ourselves to the mathematical cosmovision of the future.

On penetrating the social framework the question arises as to whether in the New Age popular resistance will disappear on contact with the higher levels of science, education and culture.

This leads us to deal first with one of the most important themes in Dr. Serge Raynaud de la Ferrière's thought, namely Mathesis. This Greek term signifies the implication of Mathema; that is to say, the categorizing in three aspects of: first, the qualitative change of the dialectic method, thesis—antithesis—synthesis—mathesis; second, the investigation of the living synthesis in leading to the integral coming into being of man as the medium of universal identification; and, finally, the naming of one of his principal works *A Mathesis of Psychology*.

As a result of this we can proceed to the farther shores of a new concept of divinity: (the Supreme Cause contains both the Non-Manifested Effect and the Manifestation), without abandoning the scientific causal basis. "Life" he says, "in its deepest sense, is a domain of precious investigations and should be appreciated in its true value; those things we call sacred are exactly those which contain the most life. LIFE—FORM—THOUGHT," he adds, "This is the point of all convergence, the final point of all investigation." This second concept is derived from a comparative study of the religions of the civilized world.

With the tacit support of Jean Piaget in the trinomial "Genesis—Structure—Knowledge", he directed his thought to the study of religion, through which arose his concept of a feasibility of uniting religions in both the collective conceptualized sense and that of individual comprehension, thus permitting the scientific establishment of a union between: (a) the objective demonstration of universal existence presented by science and, (b) the subjective understanding of the Great All, represented by a universal mysticism.

Our study continues with an investigation of other aspects of his mathematical thought such as: a) 'anticipating problems', b) the vernal point as the sensitive axis of the equinoctial precession, c) the combination and investigation of the representative values of numbers, d) his visit to the Buddhist Mathematical Society, e) his mention of "Transcendental Problems", and f) his integration of primitive and modern mathematical thought, i.e. first, the Magic Squares and the Greco-Latin Squares, second, mathematical combinations, third, chess played by computer, and fourth, the theory of games.

We have undertaken to present this manual of living, *I Realized God Through Mathematics*, in such a way as to enable it to be followed in a practical as well as a logistic manner. Some of the first questions that will arise will be the following:

What gave rise to this living example and what alternatives exist for following it?

Does it encompass only the mathematical field or does it involve scientific and social thought also?

What did Dr. Serge Raynaud de la Ferrière think about the numerical science and its mathematical surroundings, and what were his goals?

What was his mathematical and sociological viewpoint?

What was his transcendence towards sacred thoughts?

These and other similar questions may arise and for this reason we will treat the alternatives systematically.

The first alternative is through the way: mathematical models, biological structures, and structures of knowledge, as a logistical realization.

The second alternative discusses the field of analogical thought allied with concordant methods, including the reality of analogies of global structure, as in the case of archaeometry: that is to say, analogy and the universe.

The third alternative (through the epistemology of life) is, according to Dr. Serge Raynaud de la Ferrière himself, the one on which we must focus in order to analyze the different manifestations of the Supreme Cause; and the one that brings a transformation of mysticism as well as a 'consciousness of-self' of science.

In this way starting from mathematics we have once more returned to mathematics in a tour of the basic thought of Master Serge Raynaud de la Ferrière, since more than part of his scientific thought is part of his universal thought.

This has been done to enable us to establish ways and models of identification so that mankind, thinkers, and scholars may view different panoramas of the process to arrive at the living experience (*vivencia*).

We have given these processes and the framework around them, and as we state in our final paragraphs, we did not do so only for explanatory purposes, rather we were spurred on by the social function of scientific understanding, by a chart of values for a new age, that is to say, for a new humanity in which reason and knowledge will guide the highest values of human transcendence.

PART I

THE BINOMIAL:

Serge Raynaud de la Ferrière—Julian Huxley

CHAPTER 1

Two Figures in Agreement

It is interesting that the objective scope of the work of the French sage, Serge Raynaud de la Ferrière, in consonance with some outlines emanating from the English sage, Julian Huxley, virtually coincide like the action of two giant figures agreeing on what UNESCO meant to both. This leads us to come together with what brings such primordial thought to mankind.

In this analogical chapter, we are going to treat both plans by an apologetic criterion because within the books and letters of Dr. Raynaud de la Ferrière, and the books and memoirs of Julian Huxley, there exist specific points of special coincidence that help to establish more profoundly the subject's framework and above all, the good for humanity.

Perhaps UNESCO survives through many ages as an institution, but the projection of its ends and objectives in science, education, and culture has to serve as a model and incentive to organizations that, like the Universal Great Brotherhood of Dr. Raynaud de la Ferrière, have to develop through time, becoming with time more oriented toward the ends, objectives, and standards outlined by its founder.

In what age will UNESCO attain its objectives?

How long will the Universal Great Brotherhood take to achieve the standard in science, education, and culture that its founder set for it and to measure up to UNESCO?

Will the perhaps excessive governmental and bureaucratic evolution understand the necessity for developing UNESCO's high ends as far as the reality of the popular world?

Will the members of the Universal Great Brotherhood, from its perhaps too popular and mystic base, understand the organizational height desired by its founder in order to bring the institution to its highest level?

Will the respective visions of Serge Raynaud de la Ferrière and Julian Huxley be grasped quickly by history, and by the social, cultural, and spiritual evolution of people in the future?

The analysis of precessional ages, when applied to the present, indicates to us that in the measure in which the characteristics of knowing develop, the thought of Huxley, and especially the thought of Serge Raynaud de la Ferrière (above all because of his prospective determination toward a new age), will have their true world application at the high level they expected.

For a beginning let's say as a small barometer of their parallelisms, the following agreeable coincidence is curiously significant. In 1930, Huxley proposed to the English Government a plan to create various reserves destined to conserve the animals in Western Africa, but a number of years would have to pass before his proposition would be accepted. On his part, the Master in 1946 proposed to the French Government an expedition to South America for bringing about an investigation of ancient civilizations. The Minister had just finished approving the expenditure of five millions for the Charcot expedition to the Pole, and politics was of no advantage for those seeking help from the State. Both men faced similar obstacles with their respective governments of England and France.

From a viewpoint of sociological analysis, Dr. Raynaud de la Ferrière, after busying himself with the integrity of man in the application of methods, as well as thought, universality, and ontological self-realization, mastering equally the dialectical and theoretical methods, the discourse of thought, and science itself[1], would make a synthesis through the thesis-antithesis-synthesis process that would lead him to the *Mathesis*, (living synthesis) as a change which is qualitative, investigative, and experiential.

With respect to some of his investigative works, Serge Raynaud de la Ferrière said:

> "Personally, I recall the memorandum I thought to take to the French Government in order to present a project

for machines that captured solar rays, little mirror lenses capable of using the energy of the sun in the Sahara. That happened in 1938. Much later, in 1946, I thought also of solar ovens and of installing them in North Africa, at the same time making suggestions to my disciples in Venezuela about the possibility of using the forces of the sun. But I always found disbelief everywhere. But now we see enormous installations of similar projects and instruments that capture the energies of the sun, reflect and transmit them using special apparatus, etc. . . . Accordingly, with some converters of solar energy and other instruments, it will also be possible to use the chemical elements of the moon for establishing a manner of existence on this planet."

And he completes this interdisciplinary scientific prospective, by adding that:

"In what concerns the air that will be breathable oxygen and nitrogen are removed from rocks when water meets them under crystallized form in the different igneous rocks, all this calculating the percentage of evaporation with the temperatures of the solar oven. Hydrogen, carbon monoxide, carbonic gas, etc. can be produced after the dissociation of the rocks and at the same time foods can be synthesized through a reaction transforming hydrocarbons and the other elementary molecules into ethyl alcohol, etc. The production of electrical forces with the interaction of ionic ejections (the state of the rocks in rapid dissociation within the lunar vacuum), with static magnetic fields; electric currents are then produced in a connected circuit to electrodes set up perpendicularly to the ejection and the magnetic field."[2]

It is curiously significant that in August of 1976, thirty years later, NASA prepared for approval, a project for establishing colonies on the Moon, Mars, and Jupiter.

When he was Director of the Administration Committee in charge of Foreign Relations for the Institute of Archeology and Prehistory, he said, ". . . . I have presented a work on the inscriptions on a stone which existed in France, in parallel with other petroglyphs."[3] But he commented:

"It is very rare if things appear complicated when an investigator in the vanguard exposes his theories and later all this is easily seen, that the facts are proven and the things accepted."

In another paragraph, he reflected in a similar way:

". . . the interchange of ideas and immediate participation at the time always results advantageously for all."[4]

In his works he had previously given a descriptive picture of the dawning of the new age:

"After some dozen years, and more particularly, after the end of the last World War, a current of spiritual uneasiness is moving the spirit of the investigators and informing them about the direction of the more acute, serious problems of knowing: the value and reason of life and its manifestations—whatever be its forms; the value and role of human dignity; the study of the relations of men among themselves; the arousing of the intelligence, and the ethical, religious, and mystical sentiments; the difficulties, complications, and instability arising from social and economic life; and many other vital problems.

All those beings who descend within themselves report a crisis was suffered in which the spirit and its power, lethargic in the better part of men, brutally rebelled in some but more modestly in others. But each one understands that the problem has to be resolved within himself and that, only after each one has realized in himself and by himself the Truth which is perceived, will these great evils that humanity suffers be diminished and completely liquidated in a total understanding and an impartial love of rational beings among themselves and for the entire creation."[5]

"Now and then this investigative spirit manifests itself with more energy, as in the present time more than in any other time in the last two thousand years. Clearly, the social and religious manner of our civilization is at the point of dressing itself in a suit of heretofore unknown cut and color."[6]

"After these last two millennia, the more diverse philosophies have intended to reveal the current thought. The cause of this phenomenon is rooted in the fact that the dynamics of thought, upon exploring new horizons, establishes each time a new opportunity for man's development, which has to develop through the complexity of ideas of his contemporaries.[7] The time has come for each man to fulfill his true mission on Earth. Human beings ought to join together, the age of destructive criticism should come to its end, the religious or ideological divisions disappear with greater swiftness each day. And at the head will not be the politicians, but the psychologists, pedagogues, initiates, and wise men."[8]

In another part he comments with great force about hope and renewal:

"It is encouraging to observe the public becoming less rigid and more objective in its acceptance of things, more open to understanding and more ready for investigation.

" . . . it is necessary that the world understand that it is resolving the individual problem as it comes to establish an Age of Harmony, in spite of the force of politics or of governmental methods tried until this day: that is, all the lapse known by the name of history of humanity, though they existed in another age, periods and times of peace, golden ages."

Without forgetting our comprehension of the thought of the wise man, Serge Raynaud de la Ferrière, let us now proceed with the necessary generality to the wise man, Julian Huxley. With the help of the article, "Julian Huxley, Scientist, Humanist, and Thinker", written by the renowned Brazilian historian, Paulo E. de Berredo Carneiro (Director of the International Commission for preparing *History of Humanity* and the author of *About A New Humanism*), we transcribe some words of Huxley:

"That which I proposed was to achieve a better education for the unfortunate; a better understanding in all nations of the role of science, above all, of Biology and

Psychology; the exchange of students and professors between countries of different ideologies or on different levels of cultural development; the conservation of natural beauty and places of historical interest; a deeper understanding of evolution and its processes, in human society as well as in nature; the extirpation of the ideological and nationalistic interferences in questions of art, literature, and science; the adoption of methods against excessive population increase, which in this matter I divided my activities between the World Health Organization (WHO) and the economists. In spite of all this, the consciousness of the unity of humanity and of what is necessary to coordinate efforts to achieve cultural and social progress on a world scale."

Thus, Julian Huxley expressed himself in his *Memories*, referring to the program objectives of UNESCO that Huxley himself, as Secretary of the Preparatory Commission, took charge of elaborating in 1946. Paulo de Berredo Carneiro then comments:

"How few in that era were those who would have been able to assume the direction of an enterprise so audacious and so new as the United Nations Educational, Scientific, and Cultural Organization! In 1925 Huxley left Oxford and became Professor of Zoology at King's College In London. This gave, then, a new orientation to his investigations and his writings. Social problems occupied the first plane of his intellectual concerns. Huxley brought to light the scientific methods which he had practiced so often in his works.

The biologist is transformed into the sociologist, into the humanist and the educator, desirous of putting his experience and learning at the service of a restless and disoriented world. But in all of this, the man of science remains whole, but without limiting the role that belongs to the imagination in treating such complex questions.

In 1927, H.G. Wells invited Julian Huxley to participate with him and his son, G.P. Wells, in the making of a biological encyclopedia, *The Science of Life*. Huxley had to oversee the major part of the scientific work. As a result, he resigned his chair at King's College and dedicated all his time to preparing and editing the large work.

The encyclopedia was finished at the end of three years of unceasing labor. It had offered him the opportunity of enriching his natural abilities in the many subjects with which he was obliged to deal and of succeeding as a commentator because of his admirable talents.

"From then on, he resorted more each time to his spirit of synthesis for elaborating well-known scientific texts, which later would bring him the Kalinga Award, granted by UNESCO.

In the middle of such works and concerns, Huxley even found time to write one of his more important books, *Evolution, a Modern Synthesis*, which appeared in 1942 during the middle of the war and a little after his resignation as Secretary of the Zoology Society. 'It is perhaps a work most pleasing to me,' he observed in his *Memories.*

In view of Huxley's special feeling for his work on a modern synthesis, it is interesting to note the parallelism with the thought of the Master Raynaud de la Ferrière, who based a large part of his thought of a new age on the criterion of synthesis, raised to the qualitative change of mathesis (living synthesis), of which we will speak in one of the next chapters. Let us see, for the moment, some of the points of coincidence with regard to a modern mysticism.

NOTES

1. A Mathesis of Psychology. p. 17.
2. His Instructional Letters from France, by Dr. S.R. de la Ferrière. Tome III. Edic. Peru, p. 295.
3. ibid. Tome II. 191.
4. ibid. Tome III. pp. 284, 294.

5. The Great Messages, Dr. S.R. de la Ferrière. p. 385.
6. ibid. pp. 385-6.
7. ibid. p. 413.
8. Psychological Talks. Tome II. p. 61.

CHAPTER 2

Points of Coincidence With Regard to a Modern Mysticism

From our notations and comparisons, there are interesting results for the subjects we are pointing out; while at the same time, certain points of coincidence can be seen between both wise men in their concepts of a modern mysticism maintained in some manner by scientific thought.

In the work, *Religion Without Revelation*, by Julian Huxley, we are able to point out some passages which parallel and support the concept of divinity which flowers in the thought of the Master Raynaud de la Ferrière. Huxley regards it as possible to unite and harmonize men *par excellence* among themselves by means of a scientifically inspired moral system. To make his ideas known in this respect, he wrote, in 1927, a book of high philosophical value which is, if we leave aside his biological works, his most important book. Almost without exception, concludes Barredo Carneiro, the elements and practices of present religions could be utilized by a religion that would abandon the interpretation of the world based on the function of an independent God and would adopt a base of scientific concepts. As Julian Huxley expressed it:

> "This new religion ought to contribute to ending the fear of man and to freeing him from the feeling of sin. It can reveal to its followers unexpected spiritual riches, possibilities of its proper nature which they ignored. But in order that a new religion constitute a true force in human matters, it is necessary (apart from the spiritual refuge of the consolations and the occasions for doing

> good works which it offers) that it appeals seriously to the moral sense and to the imagination of humanity . . . Thus considered from either a human or a scientific viewpoint, a new religion cannot be a religion of negation, of death, of asceticism, or of resignation. It has to be a religion of life. For the job of the individual conscience consists in affirming that life can be sacred."

If Huxley sketches it, then Master Raynaud de la Ferriere, upon living it, structures it, experiences it, applies it, and enlarges it as a living and collective realization, joining it to a universality of projections. Huxley presents the beginning of a new era; the Master establishes its study, its investigation and its significances.

We appreciate this sacred sense of life in the Master when, at the end of *A Mathesis of Psychology*, he expresses his repeated phrase:

> "Life, in the most profound sense, is a domain of precious investigations which ought to be appreciated in its most exact value; the things called sacred are precisely those which have been given the most life."[1]

Julian Huxley, biologist, humanist, and sociologist, is thus joined to Raynaud de la Ferrière, mathematician, medical doctor, illuminator of an "Age of True Investigation", in his epistemological projections. His principal thesis, in *Essays of a Biologist*, notably parallels the primordial definition of Master Raynaud de la Ferrière. With respect to Life-Form-Thought, Huxley says:

> "I propose, then, to lay down as my main point that the idea of God is an inevitable product of biological evolution, arising when the human type of mind first came into being as an organism capable of generalizing, of framing concepts, and of communicating them to his fellows. The microcosm becomes more highly organized; from rough-and-tumble cinema it develops into an elaborate drama, whose plot is knotted up in the same general way as that of the great macrocosmic drama unrolling itself outside. Microcosm images macrocosm more nearly, both in its form and in its scope. As result

of this, LIFE is for the first time enabled in man's person to frame some general ideas of the outer world. Not only is it enabled, it cannot help but do so. The outer world is there; it impinges through man's sense-organs on his mind, and his mind is so constricted that, if it thinks at all, it must think in general terms. For the first time, life becomes aware of something more than a set of events; it becomes aware of a system of powers operating in events. These powers (to use a general and what is intended to be a noncommittal term) are in constant action upon man's life.

"And man frames his own idea of these powers; and once that idea is framed, it exerts an effect upon the rest of his ideas, upon his emotions, upon his conduct. The more strongly the idea is held, the greater the effect.

"The contribution to the idea of God from within, from the mind of man himself, is its FORM; and this form is the outcome of a process of mental organization occurring in the unborn embryo.

"The system of ideas which man holds concerning external powers may be thus organized by THINKING of it in terms of magic, of 'influence', manifesting itself in different ways in different operations of Nature; or in terms of personality, the manifestations of power being supposed to result from the activities of a being or beings more or less similar to ourselves; or it may be organized, as we shall see, on more scientific lines, by carefully pruning away all parts of it which are either definitely the mere product of our own imaginations, or else are not proven."[2]

Isn't the concept of the Trinity involved here in what Dr. Raynaud de la Ferrière has insisted is Life-Form-Thought? The Master explains:

"We have seen that all phenomena is Life-Form-Thought. The history of the development of this trinity: LIFE (first attribute), MATTER (second attribute), and CONSCIENCE (third attribute) is that of a long planning of the substance, assisted by the exterior Divine Life, but following the process of development of Life-

Thought-Inclusion in the generating source of the form. This Life-Thought begins to specialize when it culminates in human form, and in addition it individualizes, thus producing the human being, Conscious of Itself.[3] The irradiant force of life attains a different development according to the rules of Nature. There is a point at which those rules are confused; the Life-Inclusion, in the process of its manifestation, is guided by continuing electromagnetic waves, the morphological Laws, but there are moments in which it appears to vacillate about the route to follow. The sea anemones, the coral, the madreporas, all the polyps . . . grow and bloom like the plants; they eat, they react immediately. They extend their arms for catching their prey, like the animals . . . The family of hydrophytes or protophytes are part of these singular beings.

In the mineral state, the role of the Life-Inclusion is simple; it responds to the vibrations which it receives from the Life-Exterior. This molecular manifestation is ITS conscience. In the vegetable kingdom, development of the organs necessary for the nutritive, respiratory, and reproductive functions, will appear under the effect of a more elevated consciousness. And in the animal kingdom there appear the first vibrations of the intellect, as a consequence of the vitalization of the mental covering which produces a state of attention. And here a new function begins to be born: the differentiation which makes the Life-Inclusion attain consciousness of the exterior. The critical step of the evolution is, then, this mental plane which distinguishes animals from man. Although in the lower mind, it can be classified as a real spirit to the man who thinks only for procuring better well-being, and it must be classified, as well as the abstract spirit (or pure reason), in the superior mind with all the researchers, scientists, philosophers, etc. . . .[4]

And in the chapter, "Analysis of Life", he states:

"In his time, E. Arnaud separated three principles; a first aspect, (energetic) called Spirit-Life; a second (concentration) called Matter; and a third aspect (constructive-polarizer) called the Thought. With Energy, Matter, and Thought, he constructed a system of differentiation, ranging from Ether No. 0 (Absolute) to Ether No. 7 (the Physical World).

These three principles: Energy, Matter, and Intelligence, are like the customary triad (body, soul, spirit). Charles Henry, who also considered this problem, tried to contribute a mathematical solution. This wise man, Director of the Sorbonne Laboratory of Paris, says that life can be considered as the result of a balance between gravitational or material forces, biophysical forces, and electromagnetic forces. These forces, which he designates by the name of resonating, are capable of contributing new insights, not only to the problem of life, but also to that of survival.[5]

". . . thus there comes a day for understanding thought as a function of life and then, like a function of inert matter, will be used for discovering that life and matter have properties very different from those which were attributed to them before . . ."

In the chapter "Scientific Conclusion", he adds:

"On account of its awakening to activity, the understanding, the One is going to be the cause of all that will have to be manifested, leaving a likeness of itself. Parallel to this, the human being is able, in all things, leaving himself aside as always the same, to fulfill a potential and to realize a thought, and to be thus the cause of all which bears this fulfillment, because he possesses, from the beginning, all the possible manifestations of which he will one day be able to reveal."[6]

If we return to the chapter, "Scientific Conclusion", we find that he brings up the trilogy Life-Form-Thought in relation to the natural capabilities of a new concept of the divinity which unites religions.

"Nature is that which appears to us as joined to the phenomena of the Cosmos. Whence every phenomenon is LIFE-FORM-THOUGHT. Those which, in turn, are not without vibrations, waves, vortices, and concentrations in different ethers. The creative activity of nature is that which is present to us as the Spirit-Life (first manifestation of phenomena) of the One-Principle (third manifestation of the Thought) by means of the Forms (the second manifestation).

"All this is governed by a cyclical law: birth, development, disappearance. The duration of some cycles of propulsion varies by fractions of a second (lightning, for example) and is able to go toward the unimaginable duration of the Cosmos.

"To take another example, the Earth (like all the planets of the solar system) on account of its movement, receives daily, activity from the sun. This propulsion of energy spills over all things, according to the defined cyclic form: in the morning—the beginning; at midday—the maximum; and in the afternoon—the disappearance. Immediately, after having undergone this propulsion, a period of repose (night) comes upon it after this vitalized phase. Other activity is produced by the Earth's revolution around the sun, the change of seasons being an obvious manifestation of this, as are the tides, which are propulsions of energies that occur every 11 years here, and indefinitely in other systems. Activity-passivity, they are pulsations of energy always followed by periods of rest.

"Modern astronomers have discovered in the celestial spaces another vast cosmic pulsation at a velocity proportional to the distance; the spiral nebulae are in flight in space and this is what is called dilation of the Cosmos. You can be certain that this phenomenon will be followed by another, that of contraction, which will be given to be contemplated by another humanity.

"Raised by propulsion upon propulsion, we arrive at considering the initial propulsion that contained the Origin—the cosmic cycle ruling all the others. All these cycles come into being, develop, and disappear,[7] and the global cosmic cycle inherits the vitality originated in itself, by means of these successive propulsions.

"All in the cosmos is linked methodically, from the infinitely great to the infinitely small, in a harmony which is the Law of the Principle-One manifested, that self-fulfills the Ideation, by revealing its possibilities.

"The origin of a Cosmic Cycle is one of the most difficult things to conceive. For this, it is necessary to admit an awakening of dynamism which makes it possible for a finite Cosmos to emerge from an Infinite, Absolute Universe. This can be defined without possible attribution, if it is not from what contains all of a potentiality.

"This mystery is reproduced each day in our physical world: the dream, for example, in no way destroys our faculties. The period of passivity in which we have been submerged during the entire night conserves and even restrengthens our potential, which is going to be launched in a great manifestation at our awakening.

"On account of its awakening to activity, the understanding, the One is going to be the cause of all that will have to be manifested, leaving a likeness of itself. Parallel to this, the human being is able, in all things, leaving himself aside as always the same, to fulfill a potential and to realize a thought, and to be thus the cause of all which bears this fulfillment, because he possesses, from the beginning, all the possible manifestations of which he will one day be able to reveal."

"The manifestation of the One-Principle is a 'Lord-of-Himself', just as the infinite is going to be manifested as finite. Religions have simply taken a limitation of this manifestation under the aspect ACTIVE, which, according to the conception, is called God, or First Principle, Spirit-Life, Thought-Foundation, Essence-One, Great Architect, Energy, Cosmic, the Word, Destiny, etc. . . ."[8]

Turning to Huxley, who makes a further analysis of the restored role of science for the spiritual values we are reminded of the phrase that the Master Raynaud de la Ferrière repeats, "A little understanding draws us away from God, but enough understanding brings us to Him." Huxley points out:

"Now this view, or any view of God as a personal being, is becoming frankly untenable. The difficulty of understanding the functions of a personal ruler in a universe which the march of knowledge is showing us ever more clearly as self-ordered and self-ordering in every minutest detail is becoming more and more apparent. Either a personal God is a ruler without power, or He *is* the universe. In the former case He becomes a mere fly on the wheel; in the latter we revert to a frank pantheism, in which the idea of a personal Being can no longer properly be upheld. A personal creation of the world, in any reasonable sense of that term, is now meaningless except for a hypothetical creation of the original substance of the cosmos in the first instance. Creation of earth and stars, plants, animals, and man—

Darwin swept the last vestiges of that into the waste-paper basket of outworn imaginations, already piled high with the debris of earlier ages. After the psychological insight which the last half-century has given us, miracles have ceased to be miracles, and have become either delusions, or, more frequently, unusual phenomena for which a cause has not yet been found. The immutability of the fundamental laws of matter and motion, more particularly the grand generalization of the conservation of energy and the substitution by science of an orderly for a disorderly conception of nature, make it impossible to think of occasional interference by God with this world's affairs. Accordingly the value of petitionary prayer falls to the ground. Revelation and inspiration have resolved themselves into exceptional mental states, and are no longer looked upon as a sort of telepathy between divine and human minds. If we reflect, we see that all these intellectual difficulties in modern theology arise from the advance of scientific knowledge, which has shown that the older ideas of God were only symbolic, and therefore false when the attempt was made to give real value to them.

"That being the quagmire in which traditional Christian theology is floundering, it behooves us to discuss the opposite side of the question, and to see whether the very advance of science which has seemed to exert only a destructive influence may not have made it possible to build up new and sounder conceptions of fundamental religious ideas. We have already seen that the conception of God always represents man's idea of the powers operating in the universe; that it has two components—the outer consisting of these powers so far as they are known to man, the inner consisting in the mode in which the conception is organized and the way it is related to the rest of the personality. It is obvious that both man's knowledge of the psychic powers as well as his method of organizing them in his mind can grow and change; and man's gods can—and do—grow and change accordingly. The growth of science in the last few centuries has radically altered our knowledge of the outer world. It has shown us, in the first place, a fundamental unity of all phenomena, however apparently diverse. It has shown us the inorganic part of the cosmos pursuing a direction—the progressive degradation of energy—which, if it is carried to its limit, will result in the extinction not only of life, but of all activity. It has next shown us the organic part, sprung from the inorganic but running a different course, ascending during evolutionary time to increasing heights of complexity and to increasing control over its inorganic environment.

"Finally, we have the psychozoic or human portion—that minute fraction of the cosmos which yet is of a preponderant importance, since it definitely represents the highest level yet reached by evolutionary progress. In this sphere, mind is the dominant partner, biologically speaking, in the mind-matter partnership; evolution can begin to be conscious instead of fortuitous; and true *values* arise which, incorporated in ideals and purposes, exert an effect upon events. As regards our own mental organization, psychological science has recently shown us the enormous importance of what we may call the extra-personal portion of our mind—all that which is normally sub-conscious, or has not been during our mental growth incorporated to form an integral part of our private personality. But this extra-personal part of the mind may from time to time erupt into the personal, and does normally do so at some period of life. It is the merit of psychology to have shown the true nature of this relationship between personal and extra-personal, which was in the past a source of an infinity of mistaken ideas—revelation, inspiration, possession, direct communion with angels, saints, gods, or devils, and so forth. Thus the powers operating in the cosmos are, though unitary, yet subdivisible; and, though subdivisible, yet related. There are the vast powers of inorganic nature, neutral or hostile to man. Yet they gave birth to evolving life, whose development though blind and fortuitous, has tended in the same general direction as our own conscious desires and ideals, and so gives us an external sanction for our directional activities.

"Put broadly and roughly, there are, then three main accounts possible, or at any rate actually found in occidental civilization today, of the phenomena generally known as religious. The first is that of the out-and-out skeptic—that they are all illusions, imaginations of the childhood of the race. This is an extreme view which I do not feel called upon to discuss. The second is the view of almost every existing religious denomination in Europe—that God is a personal being. And the third is one, only just beginning to take shape, which I have endeavored, with every consciousness of inadequacy, to outline—the account made possible by a radically scientific view of the universe.

"Before the scientific work of the last three or four centuries, it was impossible to attempt what we may call a realistic account of this nature, so that symbols were perforce adopted. In Christian theology man formulated a coherent scheme, which, however, was purely symbolic, to account for the facts we have just been considering. The chief feature in any such scheme must be the conception of the powers with which man feels himself in relation; and in this particular formulation his conception of these powers was that of a God who was also a person.

"If a personal God was ruler of the universe, then he must be omnipotent; if truly divine, then omniscient; if worthy of worship, then all-wise. He must be capable of interfering with the course of events by 'miracles' of granting our prayers, of communicating directly with us, of deciding our fate in afterlife. From these conclusions yet further conclusions were drawn. If God revealed himself in the Bible, then the Bible was 'true' . . . with all that this in its turn involved as to our beliefs concerning natural causation, creation, our relations with God, or personal immortality. The whole scheme was self-consistent, and worked as well as many other human schemes. But what if the whole premise, of God as a personal being, ruler and father and judge—what if this were not in fact tenable? Then, of course, the whole edifice itself would come toppling down. That is what is actually happening today. God, as personal ruler, is being slowly driven out of the universe, but returning as this organized idea of which we have spoken."[9]

This last places us in that part of *A Mathesis of Psychology* where Dr. Raynaud de la Ferrière affirms:

"LIFE-FORM-THOUGHT. I have here the point of convergence or final limit of all investigations. To this trinity belongs: the TEM-SHU-TEFNUT of the Egyptians (in place of the Horus-Isis-Osiris myth) and the natural and manifested Creator God in the great mystery of LIFE (in place of a God), the anthropomorphic father who limits sincere conceptions; and the Form, which comes in its regular time, to materialize the Divine, like a material emanation of the Life Essence, like a true avatar (Messiah): the CHRIST, as master Jhesu (in place of a historical Jesus, constantly exposed to criticism), as a spiritual realization; and the Eternal THOUGHT, understood as a state of universal consciousness (in place of a hypothetical Holy Spirit).

"One has to succeed in overcoming the manner of restricting God with a word, (especially when its significance is so poor, for example, 'God' for the English, 'Gott' for the Germans, 'Dios' for the Spanish, 'Allah' for the Muslims, 'Bhagavan' for the Hindus, etc.). Divine existence may be denied without some attempt to do this. Likewise, one can succeed in overcoming the adoration of an unknown 'being' by realizing that we are an emanation of the Great All, a part of this Universal Body. To be unified with this Principle is to reach an intense communion with Eternity. To be unrestricted in our adoration means to be in perfect harmony, and it is not important in what temples, churches, or pagodas this occurs. For, the ESSENCE-ONE is always everywhere (without special hours). So, it is not necessary to circumscribe a place of worship, to found new religions, or to have hours specified for the ceremonies. . . . Far be it from me to criticize organized churches. I have always been a defender of order and discipline. Nevertheless, if after such a long time these have not been able to offer the immediate state of beatitude to anyone who seeks refuge in the Supreme, they have to understand that it is necessary to permit one to go from there to where one wants, when one wants.

"The diverse channels of human thought are indispensable for placing the spirit at the door that will lead it to the garden of personal inquiry. All religions have methods of moral conduct which it is necessary to respect. Perhaps the only observation that I would be permitted to make is that we have either too many or too few religions. That is, there are 'too many' religions in the sense that the sects ought to be reduced to a single RELIGION ('religare' from the Latin, to join, to unify or to bind), since the basis of all these doctrines which divide the present world is identical. Or, on the contrary, there are "too few" religions since we scarcely have three hundred today and because if each individual belonged to a different religion that would be magnificent. We would then be in the presence of a phenomenon of peculiar personal reaction, which would be the special object of psychological research. The three hundred most important religions and sects which presently exist in the world, are in turn subdivided internally, on disagreements of dogma, details of ritual, or diverse theological interpretations, thus giving an opportunity for real progress which does not compare with the pretended religious immutability. There are no immutable religions; all of them have modified their primitive dogmas. All of them have transformed a part of their teachings. Only Yoghism has remained unchanged through the centuries.

"Through my own initiative I have given the name, Yoghism, to this teaching in order to distinguish it better from Yoga about which so much has been said. I propose then, to reveal my point of view about what I call Yoghism, or the system which, by expressing all things, excludes channelizings and incomplete conceptions.

"It would please me to ignore all the conceptions, especially the traditional ones, if it were not necessary to resort to the teachings of the past to understand better the clarification which I offer, above all to the Western world, whose mentality is widely critical and whose methods of analysis have already achieved a natural progress of civilization in the 20th Century. Let us proceed then, according to the analytic method of investigation."[10]

For his part, in *Essays of a Biologist*, Huxley adds;

"This leads us on immediately to a subject of especial interest to us as rationalists—namely, the relation of religion to science and to free inquiry. But be it noted that it is not science and religion which are in conflict, but science and a particular brand of religion. The essence of science is free inquiry combined with experimental testing. The result is a body of knowledge, of fact, and explanatory theory, which can properly be regarded as established. By 'established', however, we do not mean that it is absolute or immutable—we expect addition and modification. But we also expect that, in the future as in the past, the additions and alterations will not involve the scrapping and rebuilding of the whole edifice, but that it will continue to be harmonious with itself, and to undergo a gradual evaluation. This has been so even with such marked changes as the discovery of radioactivity, the new outlook in psychology, or the rediscovery of Mendelism—the new, after apparent contradiction, has been or is being harmoniously incorporated and organized with the old.

"This in its turn implies that toleration should ever be encouraged by the scientist. Humility cannot be genuine if combined with unsupportable dogmatic assertion; and the recognition that the ideas of revelation and divine personality are such dogmatic assertions brings a whole new outlook into being.

"Putting matters in a nutshell, we can say that a system based on revelation or on the pushing of unsupported premises concerning the nature of God to their complete logical conclusions is bound to result in some degree of hostility to the pursuit of truth for its own sake; whereas a religious system basing itself on scientific method, while it must resign itself to being unable to produce a multitude (and indeed so beloved because it satisfies a lower and more primitive mode of thinking only), on the other hand can be assured that its knowledge and effectiveness will increase, and that contradictions will resolve themselves, provided that free inquiry, free speech, and tolerance are allowed and practiced. Attempts to reconcile the old formulation with the new facts and ideas, when not insincere, are doomed to failure because the premises of the two systems are different.

"In conclusion, we may perhaps point out some of the bearings of such a change. In the first place, the change in our conception of God necessitates the stressing of religious experience, as such, as against belief in particular dogma, or in the efficacy of special ritual. Secondly, it emphasizes the need for tolerance and enlightenment. The scientific view asserts not that its knowledge is absolute or complete, but that, although relative and partial, it will indubitably continue to grow harmoniously along the general lines already laid down.

"Another change wrought by the inclusion of all phenomena under one head and the banishment of the supernatural is the inestimable advantage that we thereby find the possibility of constructing a single general view of the universe for civilization. At present there are two that matter—the orthodox religious and the scientific. The religious starts from the top, the scientific from the bottom; but the scientific has been creeping up, and now that it has begun to attack the problem of mind it will be able to drown the other out. Since the current religious formulation is only symbolic, it cannot become scientific; but since the scientific is based on the closest possible analysis of reality, it can become religious so far as it investigates the realities of religious experience. Once it has done this, we shall be able to construct a 'Weltanschauung' such as never before, with roots in the ordered reactions of inorganic matter, trunk strong with the steady progress of evolving life, and branches reaching up into the highest realities of the spirit. Union is strength; and it is one of the prime duties of educated men and women to see that the present duality and antagonism at the heart of what should be the central unity of civilization—of its fundamental idea, its conception of the universe—should be terminated."[11]

In turn, when the Master Raynaud de la Ferrière makes an analysis of INRI he announces in the new age the union of the sciences and religions as one of the fundamental ends in collaboration with the Institution by the founder; and through which it becomes fitting that we analyze his position and ideology in comparison with the organizations which he founded.

NOTES

1. A Mathesis of Psychology. p. 503/6.
2. Julian Huxley, Essays of a Biologist. pp. 208-210, 213.
3. The Great Messages, by Dr. S.R. de la Ferrière. p. 420.
4. ibid. pp. 421-2.
5. ibid. pp. 295-6, 418.
6. ibid. p. 418.
7. ibid. p. 417.
8. ibid. p. 418.
9. Essays of a Biologist, pp. 217-220.
10. A Mathesis of Psychology, by Dr. S.R. de la Ferrière. pp. 23-5.
11. Essays of a Biologist, pp. 222-4.

CHAPTER 3

UNESCO and The Universal Great Brotherhood: The Projections of Two Thinkers

It is true that Julian Huxley was not the sole founder of UNESCO as Serge Raynaud de la Ferrière was of the Universal Great Brotherhood, but each was founded with the ideology of each one of them, respectively.

What did they think about their great foundations? How did they plan and experience the concrete realization which historically resulted from their vast coincidences?

In the beginning of 1945 Julian Huxley was named Secretary of the Preparatory Commission of UNESCO. As for Huxley, he says (in the article "UNESCO's Early Years" which was based on an interview with him in 1970 and published in the March 1976 issue of The UNESCO *Courier*):

> "When I entered the Commission's office, I felt, as I wrote in my diary at the time, like the headmaster of a school who is also its youngest pupil. After settling down, I took a fortnight off to write a pamphlet entitled: "UNESCO, It's Purpose and Its Philosophy", to clarify my own thinking and, I hoped, that of the delegates in the forthcoming first General Conference. This was to be held in Paris in December 1946 and would decide the main outlines of UNESCO's policy and programme and also elect its first Director-General.
>
> "As it turned out the humanist attitude which I adopted in this pamphlet led various delegates, quite erroneous-

ly, to think that I was anti-religious, while my liberal views were taken by others as Communist.

"We started work in January 1947 in the Hotel Majestic, which had been the Nazis's Paris Headquarters during the Second World War. Ironically enough, I, as Director-General of an organization dedicated to peace, the promotion of knowledge and culture, occupied the room which had harbored the very unpeaceful, repressive Nazi Commandant.

"The preamble of UNESCO's constitution, beautifully written by the American scholar-poet Archibald MacLeish, proclaimed that UNESCO was primarily dedicated to peace—though through education, science, and culture—and I had often to defend projects that were not specifically concerned with peace by maintaining that the promotion of education, science, and culture, especially when involving the exchange of ideas and personnel between different regions was UNESCO's primary task and could and would indirectly lead to peace through better mutual understanding.

"In education, UNESCO's first job was obviously to promote literacy in a largely illiterate world, but we soon found that literacy in the customary sense was not enough. Besides learning to read, write and do simple arithmetic, the large mass of underprivileged people needed to be educated in hygiene, in improved methods of agriculture and better care of their environment. This comprehensive program we called 'Fundamental education'. It proved very valuable in our early stages by lifting underprivileged peoples to a decent minimum of intellectual and physical advance.

"Later the phrase 'fundamental education' was dropped and there were separate projects for health, in conjunction with the World Health Organization, for general education, for science, for population control, for exchange of persons and for conservation.

"Besides this, there were numerous non-governmental associations concerned with different aspects of UNESCO's work. Some were already in existence before

1945, like the International Museums Office, which I, as Secretary to the Preparatory Commission, agreed should be affiliated to UNESCO, helping UNESCO with advice, benefiting by grants or visits by specialists financed by UNESCO.

"Others had to be deliberately created, sometimes in the face of considerable opposition. This was the case with the International Union for the Conservation of Nature, IUCN for short, consisting of specialists in ecology, in establishing reserves to protect unspoiled scenery, rare and interesting plants and animals, and in preventing pollution and unplanned urban spread.

"Many delegates to the Conference, which we held at Fontainebleau in 1948, thought that preserving wild life and the like was outside UNESCO's province, as it did not specifically promote peace. But, eventually, the conference agreed to the formation of the Union and its affiliation with UNESCO. I may add that the Union, both on its own and with grants from or personnel financed by UNESCO has achieved a great deal in preserving much of the world's beauty and its interesting plant and animal inhabitants, though there is still much to be done.

"As regards science, we found that the scientists in one country in a given region, like the Middle East, were often more ignorant of the work of scientists in other countries of their own region than of the work of leaders of science in more developed countries. The Regional Centers for Science established by UNESCO put local scientists in touch with each other so that they could apportion research in relation to the areas' need, and they also put them in touch with world science by the information we provided.

"We also set up a scheme for the exchange of individual experts; post-graduate students to work in the scientifically more advanced countries; specialists to advise on projects in so-called underdeveloped nations; exhibitions in world centers for artists in different regions, visits by eminent lecturers, and so forth.

"At the end of the Second World War, one of the tasks which had confronted the newly created UNESCO was to try to solve the problems, at times dramatic, which set forth the situation of the children in the countries affected by the conflict. With such an end, the organization promoted the accelerated construction of schools, the normalization of Braille writing, the preparation and diffusion of handbooks, etc.

"In visual arts, we set up two special projects. One was a catalogue of good reproductions of well-known paintings with miniature colored photographs of the originals and prices of the reproduction. This has proven extremely useful in schools in many countries. The other was the publication of large folio volumes illustrating little known works of painting and architecture and sculpture, such as the cave paintings of Ceylon and India. These were more valuable to specialists but certainly disseminated knowledge and appreciation which was otherwise unavailable.

"I badly wanted to set up a powerful organization concerned with population control in association with the United Nations and, I hoped, its other Specialized Agencies; but religious and nationalist prejudices permitted only a handful of experimental projects. Today, I'm thankful to say, not only UNESCO, but the U.N. itself, the World Health Organization and also the World Bank are helping in this, in my opinion, the most important task facing mankind.

"I must mention one final project suggested to me by Joseph Needham: the writing and publishing of a world history which would stress advances in science and literature, in culture and the arts, rather than political changes and military events. This was adopted, though against considerable opposition, by the UNESCO General Conference in Beirut (Lebanon) in 1948, when my career as Director-General ended. However, in 1949, Professor Paulo de Barredo Carneiro of Brazil was appointed President and myself as Vice-President of the Commission appointed to supervise the writing of the work. Today, the six volumes of the *Cultural and Scien-*

tific History of Mankind, bringing it right up to date, have been published both in French and English, and numerous other translations have been arranged.

"I can truthfully say that my two and a half years at UNESCO were the high point of my life, a time when all my capacities and all my knowledge and energies were fully involved."[1]

If, for Julian Huxley, the experience at UNESCO meant the zenith of his existence; for Serge Raynaud de la Ferrière UNESCO also expressed completely the motivation for his highest purpose of synthesis. He makes many varied references to UNESCO. For its textual value, we are reproducing this paragraph from his Psychological Talks, XIX:

"Finally, let us simply say in conclusion that the UNESCO project for international cooperation in scientific research corresponds exactly to the activities displayed by the Universal Great Brotherhood since 1948 (and professed by its founder Dr. S.R. de la Ferrière)."[2]

In this paragraph from his Psychological Talks, Tome XIX, he observes that 'he professes', and he uses the term 'profess' with special emphasis in order to make clear that the activities displayed since 1948 by the Universal Great Brotherhood, the institution which he founded for international cooperation in scientific research, correspond exactly to the UNESCO plan. And he adds that the Universal Great Brotherhood is committed to a plan of reeducating humanity in a universal cultural sense that aims at peace in the human community as well as in each individual. Dr. Raynaud de la Ferrière, in Psychological Talk VIII: The Art of Healing, devotes his whole preface to an article from the review "The UNESCO Courier" (July-August 1956) signed by Alfred Metraux. In the beginning of the same work's introduction, he exclaims:

"We shall never know how to sufficiently encourage research workers, and the curious as well, to read The UNESCO Courier.[3]

"I was interviewed in the General Headquarters of UNESCO in Paris. It was agreed that each Executive

> Council of the Universal Great Brotherhood should establish contact with the national delegation of UNESCO in its own district. As a result, the possibilities of collaboration between our institution and UNESCO, depend upon our own UGB representatives in each country. UNESCO's General Administration in Paris accepts cooperation from international organizations, but these are always French organizations, or with representation in France; however, the General Administration of UNESCO does not have anything to do with the national delegations of UNESCO in each country, which depend solely on the respective government of each country. As can be seen, it deals with independent branches of UNESCO, and not with a homogeneous whole. The national delegations of UNESCO are in themselves autonomous, this being the reason why contacts must be established with them nationally. This is difficult for us as an international institution to understand. Thus, UNESCO's plan develops on a local level, while ours develops on a world level."[4]

And he proposed as one of the most important dispositions for the Universal Great Brotherhood a high level of synthesis between the two institutions.

> "But our programs are more or less identical, and for this reason we should try and find a method of synthesis for reaching a close collaboration.
>
> "For two years I have been periodically sending important documentation (press clippings, various aspects of information about the Universal Great Brotherhood's activities, names and addresses of our representatives, events, work accomplished, etc.) to the General Headquarters of UNESCO. Mr. Zuckerman, Chief of Public Relations and Information, has just written me to thank us for so many valuable contributions, especially in the field of education, schools for children, help in behalf of literacy, etc. (those works in which UNESCO is particularly interested) and also sent me brief documentation in that regard. I immediately sent this information to the different UGB Executive Councils so that they

> might take note and use it as a model in developing our educational working plans. Mr. Zuckerman informs me also that it would be advisable for our representatives to make contact with the UNESCO delegate for Latin America."[5]

He insists on collaboration with UNESCO, which later will be one of his great preoccupations:

> "A closer collaboration with UNESCO in each country would be useful and I ask you to ascertain possible relationships. I sent certain sectors a photocopy of the credentials letter (ODG/NGO/795.540) which I received on September 23, 1958 from Vladimir Hercik, Head of the Division of Relations for International Organizations. This document confirms the registration of the UGB in UNESCO.[6]
>
> "It is not sufficient to do it here in France only. What is important is to give an account of our registration and of Dr. S.R. de la Ferrière's foundation with its various activities and to obtain in addition to the registration of each district, an efficient collaboration. Since each country has its particular activity to develop, we must offer our services and cooperate more."[7]

Two years later he addresses in a most urgent manner a few national sections that had been lazy, among the many sections operating in the fifteen countries of the Americas.

> "For eighteen months I have been waiting for the questionnaires enclosed in my Letter, Circular No. 11. Clearly, until I have answering letters in my possession, we are not able to cooperate with UNESCO, since I have to present something concrete to this international organization about our movement in the different countries where it is established. I should have at least a chart full of important data (already explained in my Circular No. 17). As it is, I'm unable to present UNESCO with complete information about our movement such as the number of Universal Great Brotherhood members in each country, the number of sectors, their activities in each one, the possibilities, the number of books and

leaflets or publications we can get, the radio stations, press and other things at our disposal for operating in the world.

"I would like to visit the different sectors incognito and anonymously to see what they can offer me. I am sure they would be unable to interest me in the movement ... it is unthinkable that a founder not recognize his own foundation and yet that is the way it is.[8]

"The news from Nicaragua is much more favorable; they inform me of a rarely attained event: the acceptance of the Vice President of Public Education as the representative of the Nicaraguan UGB at UNESCO's grand reunion in Paris. The nomination was confirmed by Dr. de la Ferrière who immediately wrote to UNESCO's General Administration in Paris to support this event at which a governmental figure is delegated at the same time to speak for our institution.

"Our Executive Council from Nicaragua published in *La Noticia* on February 1, 1959 a photograph of the 10th Grand Reunion of UNESCO in Paris. The official UGB delegation appears under the title, 'The Universal Great Brotherhood was represented in Paris by the Vice-Minister of Public Education ...'[9]

"These two examples show the different way the UGB is judged: by some people it is misunderstood; by others our movement is seen as it really is—a beneficial work of great transcendency.

"The fact that Dr. Pedro J. Quintanilla (Vice-Minister of Public Education) is once again the Nicaraguan representative at UNESCO's 11th General Conference in Paris and will at the same time gladly represent us, is good news.[10]

"Therefore, how can some say they are disciples of the Master without following and accepting the first principles?

"It doesn't seem to me that I have concealed my conceptions; they have been widely expounded in all my writings. And I should like to see them disseminated, in order to produce a change in the narrow minds of the majority of men. The New Era must manifest itself with new principles and it is a shame to see those interested in such an ideal not putting the ideas into practice, but following with definite enthusiasm the sectarian concepts that have prevented the development of mankind for centuries. It is time for a change. And the members of the foundation of Dr. Serge Raynaud de la Ferrière should be the first ones to effect this change.

"We must, by the way, once again congratulate our dear Gelong, the Licentiate David Ferriz and his collaborators. With the assistance of two hundred persons, they organized a Missionary Assembly for the purpose of discussing the fundamental tasks, assumptions, and suggestions about new activities. Their conclusions principally concern the creation of a Higher Education Committee, creation of medicine chests, dispensaries, kiosks, sewing rooms, etc. in order to obtain official recognition of the UGB as an institution of PUBLIC SERVICE.[11]

We see that he also encourages other members:

"This month I must in particular congratulate you not only for the report delivered to the Minister of Education—which reveals our interest in helping the government—but in particular for producing the leaflet which gives a good idea of what the UGB is and details especially the work accomplished. The list of the many things done on behalf of our work is truly a great thing for the name of our movement. It is a pity that all the other sections do not do a similar job. I am going to send some of these leaflets as documentation to the different centers that want to follow your example and publish similar reports of their activities.

"Forgive me for not having answered immediately and in particular, but I really have been almost 'submerged' with letters lately, so much so that it was impossible for me to answer all of them. Clearly, your missionary activities interested me a lot and the UNESCO administration in Paris is already informed of your accomplishments; in particular, I sent them press articles, principally the one that pleased me with the following title: 'The Universal Great Brotherhood organized a school for 100 children in Boca de Aroa', as well as the other article with the beautiful photograph, 'Special Commissioner of UGB created a school in Boca de Aroa'. These examples should be followed in all sectors. Thank you also for the photographs.

"You can always use my name since I am always in close contact with the Paris administration (particularly with Vladimir Hercik, Head of the Division of International Organizations). I have always sent abundant documentation about our activities in different countries."[12]

On another occasion, he wrote me in a note that at times he worked 23 hours a day and that he was writing rather than typing so as not to disturb his neighbors, adding in another paragraph that he was delighted to be receiving good perspectives:

"Surely the UNESCO question must not be left aside and I should like to see your different activities in that respect. Your correspondence and the reports sent to UNESCO and the authorities are really magnificent examples for making known the different activities carried out by your sector; it is necessary to continue without relaxing because we are at the brink of proving that our work comes before all profit for humanity.

"Certainly the UGB is honored by the offer of your systems and we are going to study practical methods for bringing them into public light. I understood well your plan of action for the social betterment of peoples and there is no doubt but that your ideas enter the framework of principles developed by our institution."[13]

Let us now complete the picture of the institutional projection of Dr. Serge Raynaud de la Ferrière with respect to that other institution founded by him, the International Federation of Scientific Societies (FISS) which is expressed in the following terms:

"In scientific spheres as in para-scientific ones, methods of synthesis are used for the purpose of adapting new disciplines of investigations.

"Twenty years ago, I personally attracted the attention of my contemporaries on the necessity of methodizing investigative programs in the various fields of science and heralded the centralization of the works of wise men. It is with such a motive that I founded in February 1947, the International Federation of Scientific Societies under the honorary presidency of General Durocq (of the French Air Ministry) and Professor Toutain (Director of Studies, the Sorbonne), with Secretary-General Ingeniero Betelgeuse, and the Secretary-Delegate Senor Rayet (Meteorologist). The Scientific Bureau was composed of the following persons: Abbott Moreaux (Director of the Astronomical Observatory of Bourges); Professor Prunier (Physicist); Mrs. Cluzel (Geographer); Professor Varcolier (Mathematician); Mr. Varhardt (Technician); Dr. Sauvageot (Medical Doctor); Professor Tranin (Ethnologist); Professor Boutaric (Biologist); Professor Ananoff (Astronauticist); and Professor Belgodere (Chemist).

"In 1960 the FISS was legally registered in the Prefecture of Nice (Maritime Alps—France) as a scientific headquarters for joining the various branches of science (Document 4891 No. 9742): Life President—Dr. S. Raynaud de la Ferrière; Honorary President—Count Gautier-Vignal (of the Mediterranean University Center); Secretary General—Professor Burkhal (Pre-Historian). The Scientific Bureau was composed of eminent personalities like Licentiate P. Isnard (Zoology), Commandant Octobon (Archeology), Mr. R. Gaillard (Biology), Mr. Thirion (of the Museum of Nice), Mr. Vindry (of the Museum of Grasse); for Ethnology, Mr. P. Couteau (of the Astronomical Observatory of Nice); Mr. Dujardin (Entomologist), Mr. Valentin (History), Mr. Gallant (Geology), and Mr. L. Rousseau (Antiquity and Literature).

"Besides the sympathizing members and the active members, there are more than two hundred titular members. Among them are such eminent personalities as the great world renowned sage, Professor Jean Rostand (Member of the French Academy and the Institute of France); the Doctor J.E. Caffort (Doctor of Medicine of the Faculty of Paris); Dr. P. Gerbaut (Director of the Center of Researches and Studies in Social Sciences and Humanities—France); Dr. Sylvain, France (from UNESCO); Dr. Ferrándiz (of the Official Medical College of Barcelona, Spain); Dr. F. Hopkins (President of the International Federation of Psychologists of England); Licentiate Napoleón Rodríguez (Rector of the University of El Salvador); Dr. Manuel Luis Escamilla (Dean of the Faculty of El Salvador); Dra. Doris de Stone (of the National University of Costa Rica); Professor P. Marchesseau (of the Faculty of France); Mrs. Olga Ilharreborde (President of the Chilean Red Cross); Professor Claudio Urrutia (of the Astronomical Observatory of Guatemala); Professor Fernando Nobre (of Brazil); Professor Kamienski (of the Polish Academy of Science); and Professor Lavachery (Perpetual Secretary of the Royal Academy of Belgium).

"The FISS now includes 68 societies (observatories, universities, scientific groups, research institutes and such) all legally registered (on stamped papers) and officially registered in the archives of the government of France.

"All the details and available information are given in a pamphlet entitled, 'A World Program of Scientific Synthesis'.

"This exposition appears to be a little outside of the framework of my Circular, but in reality it answers a little the questions that have been addressed to me on different occasions, since as President-Director General of the UGB, I am at the same time Life President of the FISS.[14]

"Nevertheless, the members of the UGB ARE NOT, for that reason, therefore members of FISS, although some adherents of the UGB are registered also in FISS like Dr. J. Pantaleone (Physician from Trenton, USA), Dr. F.W. Wosley (Director of the University of Burnaby, Canada), Dr. H. Cohen (American International Academy of the USA), Professor Uba Sein (Director of the University of

Rangoon—Burma), Professor T. Viana (of Honduras), Dr. Manuel Cabeza (of Colombia), Professor P. Gatty (of El Salvador), Dr. Boltwood (of St. Andrew's University of England), Professor John Zitko (of the World University Roundtable, USA). Many other personalities are at the same time members of both our organizations, but nomination is not automatic since it depends on a study of the particular case. It is the same for the dignitaries of UGB; they ARE NOT automatically registered in FISS."[15]

Having mentioned this other important scientific project which we have thoroughly described elsewhere, let us consider in the next chapter the ideology of Dr. Raynaud de la Ferrière as compared to Julian Huxley. This will contribute information for analyzing the framework of his sociological projection.

NOTES

1. Julian Huxley, "UNESCO's Early Years"; UNESCO Courier, Vol. 29, March, 1976. pp. 4-6.
2. Psychological Talks, XIX. p. 9.
3. Psychological Talks VII. Introduction, p. 81.
4. His Instructional Letters, Tome III. p. 146.
5. ibid. pp. 146-7.
6. ibid. Tome II. p. 158.
7. His Instructional Letters, Tome I. p. 55.
8. ibid. pp. 201, 83-84.
9. ibid. Tome II. pp. 2, 234.
10. ibid. Tome III. p. 51.
11. ibid. Tome II. pp. 2, 109, 234.
12. ibid. pp. 42, 54-5.
13. ibid. Tome I. pp. 127, 143.
14. ibid. Tome II. pp. 290-1.
15. ibid. Tome III. p. 292.

CHAPTER 4

Their Ideologies Compared

In 1946 a little before the founding of UNESCO, Julian Huxley wrote an important text titled, *UNESCO: Its Purpose and Philosophy*. The UNESCO Courier of March 1976 commented that the controversy raised by this document, by its supposed "anti-religious tendency" was such that many on the Preparatory Commission at the first General Conference of the organization wouldn't approve it. In the leaflet mentioned, Huxley said:

"UNESCO must foster and promote all aspects of education, science, and culture, in the widest sense of those words.

"The wide diffusion of culture, and the education of humanity for justice and liberty and peace, are indispensable to the dignity of man and constitute a sacred duty which all nations must fulfill in a spirit of mutual assistance and concern.

"A peace based exclusively upon the political and economic arrangements of governments would be inadequate, since it could not secure the unanimous, lasting and sincere support of the peoples of the world. And peace, if it is not to fail, must therefore be founded upon the intellectual and moral solidarity of mankind.

"The states affirming the Constitution will be able to declare their full confidence in securing, for all, the full and equal access to education, the possibility of free investigation of the objective truth, and the free exchange of ideas and understanding.

"From acceptance of certain principles or philosophies UNESCO is obviously barred. Thus it cannot base its outlook on one of the competing religions of the world as against the others. Neither can it espouse one of the politico-economic doctrines competing in the world today to the exclusion of others. It cannot do it, in part, because all sectarianism goes against the Charter and its own essence, and in part also, for the simple practical reason that all attempts of this nature would immediately unleash the active hostility of important and influential groups would involve a lack of deliberate cooperation on the part of a certain number of Member States, to whom it would include encouragement to withdraw from the organization.

"For somewhat similar reasons it cannot base itself exclusively on any essentially sectarian philosophy or restricted outlook.

"Finally, emphasizing the concrete commitments which ought to be accomplished in the matter of education, science, and culture, and insisting on the need of establishing a mutual understanding between peoples and of implanting on this planet peace and wellbeing, it does not seem that UNESCO can adopt either an attitude based exclusively or essentially on the faith in life, nor, in truth, to take as a point of departure any type of dualism. UNESCO's main concern is with peace and security and with human welfare, in so far as they can be subserved by the educational and scientific and cultural relations of the peoples of the world. Accordingly its outlook must, it seems, be based on some form of humanism. Further, that humanism must clearly be a world humanism, both in the sense of seeking to bring in all the peoples of the world, and of treating all peoples and all individuals within each people as equals in terms of human dignity, mutual respect, and educational opportunity.

"It must also be a scientific humanism, in the sense that the application of science provides most of the material basis for human culture, and also that the practice and understanding of science needs to be integrated with that of other human activities.

"It cannot, however, be materialistic, but must embrace the spiritual and mental as well as the material aspects of existence, and must attempt to do so on a truly monistic, unitary philosophy basis.

"Finally it must be an evolutionary as opposed to a static or ideal humanism. It is essential for UNESCO to adopt an evolutionary approach. If it does not do so, its philosophy will be a false one, its humanism at best partial, at worst misleading.

"The evolutionary point of view adapts the relationship between natural sciences and human history: it teaches us the necessity of thinking in a dynamic manner, on the limits of velocity and direction, and not statically, in accord with a momentary position and with a quantitative result: not only does it make us see the origin and shows us the biological roots of human values, but moreover, in the measure of the apparently neutral mass of natural phenomena, it permits giving these values certain foundations and certain exterior criteria. The evolutionary point of view is indispensable, in the measure in which it permits us to choose, in the case of tendencies which are counterpoised in the real world, the principles, the activities, and the methods which UNESCO ought to emphasize and support.

"It seems, then, that the general philosophy of UNESCO should be a scientific world humanism, global in extent and evolutionary in background.

"Of special importance in man's evaluation of his own position in the cosmic scheme and of his further destiny is the fact that he is the heir, and indeed the sole heir, of evolutionary progress to date. When he asserts that he is the highest type of organism, he is not being guilty of anthropocentric vanity, but is enunciating a biological fact. Furthermore, he is not merely the sole heir of past evolutionary progress, but the sole trustee for any that may be achieved in the future. From the evolutionary point of view, the destiny of man may be summed up very simply: it is to realize the maximum progress in the minimum time. That is why the philosophy of UNESCO must have an evolutionary background, and why the concept of progress cannot but occupy a central position in that philosophy.

"Taking a new turn, we perceive the necessity that UNESCO have a program which is a horse of many fields: in the particular case to which we refer, this program ought to involve, besides the application of medical sciences, the realization of studies on agricultural productivity (erosion, mechanization, etc.) and on social assistance.

"The key to man's advance, the distinctive method which has made evolutionary progress in the human sector so much more rapid than in the biological and has given it higher and more satisfying goals, is the fact of cumulative tradition, the existence of a common pool of ideas which is self-perpetuating and itself capable of evolving. And this fact has had the immediate consequence of making types of social organization the main factor in human progress or at least its limiting framework.

"The more united man's tradition becomes, the more rapid will be the possibility of progress; several separate or competing or even mutually hostile pools of tradition cannot possibly be so efficient as a single pool common to all mankind.

"Well now, the political unity of the world is unfortunately a remote ideal, and in any case does not apply to UNESCO. With that I don't want to say that it cannot do much for peace and security. Concretely, in its program of education it can relieve the final necessity of world political unity and familiarize all peoples with the consequences which would involve the transference of the full sovereignty of the different nations to a world organization. Above all, it can help the peoples to be mutually understood, and to understand their common humanity, their common task in contraposition to the nationalistic distinctions which they have to isolate them.

"Strictly speaking, the human individual lacks signification if he is considered isolated; he acquires direction from a determined form of society. His development is found conditioned by the society in which he has been born, and by the social traditions which he has inherited, and the value of the work he does in life depends on the social framework which benefits by it or transmits it to a later time.

"Thus UNESCO's activities, while concerned primarily with providing richer development and fuller satisfactions for the individual, must always be undertaken in a social context; and many of its specific tasks will be concerned with the means towards this general end—the improvement of social mechanisms or agencies, such as educational systems, research organizations, art centers, the press, and so forth.

"Thus in the past the great religions unified the thoughts and attitudes of large regions of the earth's surface; and in recent times science, both directly through its ideas and indirectly through its applications in shrinking the globe, has been a powerful factor in directing men's thoughts to the possibilities of, and the need for, full world unity.

"Later, its object will also have to be the elaboration of a unified and common conception and a series of shared objectives.

"This will be the latest part of the task of unifying the world mind; but UNESCO must not neglect it while engaged on the easier jobs, like that of promoting a single pool of scientific knowledge and effort.

"From this global aim, another principle immediately follows. It is that UNESCO should devote special attention to increasing the levelling up of educational, scientific and cultural facilities in all backward sectors where these are below the average, whether these be geographical regions or underprivileged sections of a population. To use another metaphor, it must attempt to let in light on the world's dark areas. The reason for this is plain. For one thing it will be impossible for humanity to acquire a common outlook if large sections of it are the illiterate inhabitants of a mental world entirely different from that in which a fully educated man can have his being, a world of superstition and petty tribalism in place of one of scientific advance and possible unity.

"Thus, mass campaigns against illiteracy, and for a common fundamental education, must form part of UNESCO's program.

"Further, a satisfactory common scale of values can obviously not be attained so long as large sections of mankind are preoccupied with the bare material and physiological needs of food, shelter, and health.

"Again, science will not achieve its optimum rate of advance, either in research or in its application, until its light is more evenly shed over the dark surface of the world's ignorance, so as to provide a more equable distribution of scientists, of apparatus, and (equally important in the long run) of popular understanding of science.

"With art and the appreciation of beauty, much of the 'dark area' is differently situated—in the centers of industry and among the proletariat of industrially advanced sections. But the task of lighting the dark spots in this field is no less urgent than in education or in science.

"Finally we come to a difficult problem—that of discovering how we can reconcile our principle of human equality with the biological fact of human inequality. Perhaps the problem is not so difficult as it appears when stated in this paradoxical form; for the contradiction largely disappears as soon as it is realized that equality is used in two very different senses.

"The democratic principle of equality, which is also UNESCO's, is a principle of equality of opportunity—that human beings should be equal before the law, should have equal opportunities for education, for making a living, for freedom of expression and movement and thought. The biological absence of equality, on the other hand, concerns the natural endowments of man and the fact of genetic difference in regard to them.

"Concretely, genetic human inequality is of two types. First, there is the inequality of mere difference. Some people are fair, others dark; some are tall and thin, others short and stocky; some have a natural gift for music, others for athletics; some are introspective, others practical and extrovertive. Indeed, we can now definitely state that no two human beings, with the single exception of the members of pairs of identical twins, are biologically equal in the sense of possessing the same genetic constitution, so that biological difference is, for all practical purposes, universal.

"Furthermore, the range and degree of genetic variety in man is greater than that found in any other animal species. This is largely due to one of man's biological peculiarities, namely that his local differentiation into races is not continued to the stage of separate and intersterile species, as in almost all other organisms, but has always been followed by migration and interbreeding.

"Secondly, there is difference in quality or level. Human beings are not equal in respect of various desirable qualities. Some are strong, others weak; some healthy, others chronic invalids; some bright, others dull; some of high, others of low intelligence; some mathematically gifted, others very much the reverse; some kind and good, others cruel and selfish.

"It is usually not so easy to say how much of this second sort of inequality is due to heredity, and therefore relevant for our purposes, and how much is due only to the effects of physical or social environment. But in most cases we now know, and in almost all can be reasonably sure, that at least some of the difference is genetic.

"These are certain, for instance, length of life, physical strength, and, most important for our purpose, intellectual gifts—both special ones like mathematical aptitude and general ones like intelligence; while it is highly probable for some aspects of moral qualities, though the situation here is more complex.

"It is therefore of the greatest importance to preserve human variety; all attempts at reducing it, whether by attempting to obtain greater 'purity' and therefore uniformity within a so-called race or a national group, or by attempting to exterminate any of the broad racial groups which give our species its major variety, are scientifically incorrect and opposed to long-run human progress. On the contrary, UNESCO should aim at securing the fullest contribution to the common pool from racial groups which, owing to their remoteness or their backwardness, have so far had little share in it.

"To adjust the principle of democratic equality to the fact of biological inequality is a major task for the world, and one which will grow increasingly more urgent as we make progress towards realizing equality of opportunity. To promote this adjustment, a great deal of education of the general public will be needed as well as much new research; and in both these tasks UNESCO can and should cooperate.

"The task before UNESCO is necessary, is opportune, and in spite of all multiplicity of detail, is single.

"That task is to help the emergence of a single world culture, with its own philosophy and background of ideas, and with its own broad purpose. This is opportune, since this is the first time in history that the scaffolding and the mechanisms for world unification have become available, and also the first time that man has had the means (in the shape of scientific discovery and its applications) of laying a world-wide foundation for the minimum physical welfare of the entire human species. And it is necessary, for at the moment two opposing philosophies of life confront each other from the West and from the East, and not only impede the achievement of unity but threaten to become the foci of actual conflict.

"You may categorize the two philosophies as two supernationalisms; or as the American versus the Russian way of life; or as capitalism versus communism; or as Christianity versus Marxism; or in half a dozen other ways. The fact of their opposition remains and the further fact that around each of them are crystallizing the lives and thoughts and political aspirations of hundreds of millions of human beings.

"Can this conflict be avoided, these opposites be reconciled, this antithesis be resolved in a higher synthesis? I believe not only that this can happen, but that, through the inexorable dialectic of evolution, it must happen—only I do not know whether it will happen before or after another war. Since another war would be so appalling as to set back the march of human progress by centuries, I am convinced that the task of achieving this synthesis in time to forestall open conflict must be the overriding aim of UNESCO.

"In pursuing this aim we must eschew dogma—whether it be theological dogma or Marxist dogma or philosophical or any other form of dogma. East and West will not agree on a basis for the future if they merely hurl at each other the fixed ideas of the past. For that is what dogmas are—the crystallizations of some dominant system of thought of a particular epoch. A dogma may, of course, crystallize tried and valid experience; but if it be dogma, it does so in a way which is rigid, uncompromising and intolerant. What, for wanting a better term, I have called doctrine may also embody valid experience; but it may be flexible, may be capable of growth and development and adjustment.

"Some dogmas may represent a more recent past than others; but that does not render them any the less rigid and accordingly any less dangerously out of date, any less incapable of reconciliation with opposing systems. If we are to achieve progress, we must learn to uncrystallize our dogmas.

"The two opposing philosophies of today differ essentially on one point—the relation between the individual and the community. But this one central difference provides differences in every field with which UNESCO has to deal, as well as in many others.

"It engenders different moralities and systems of ethics; different methods of education; different conceptions of the role of art in society; different economic systems; different ways of integrating science with national life; different interpretations of the fundamental human freedoms; different conceptions of the possibilities and limits of international cooperation.

"I believe that these differences, though they will undoubtedly become irreconcilable without armed conflict if they are permitted to express themselves as dogmas, to embody themselves in rigid social systems, and to become translated into terms of politics and power, can in principle be reconciled. They can be reconciled along the lines of some such evolutionary humanism as I have sketched in my opening sections, in which, though the full development of the individual is recognized as the central aim and criterion of further evolutionary progress, the proper organization of society is recognized as the indispensable mechanism of that progress.

"Put in another way, society as such embodies no values comparable to those embodied in individuals. But individuals are meaningless except in relation to the community (though that community transcends the nation both in space and in time), and can only achieve fullest self-development by self-transcendence, by interpenetration of the self with other reality, including other selves.

"The problem is thus not one of metaphysics or of dogma, but essentially practical—how best to adjust or still better to reconcile the claims of two concrete sets—individual human beings, and human social organizations.

"Accordingly, I believe that this reconciliation can be approached from two directions. It can be approached from above and from outside, as an intellectual problem, a question of agreement in principle; and it can also be approached from below and from within, as a practical problem, a question of agreement through action.

"The world is potentially one, and human needs are the same in every part of it—to understand it, to control it, to enjoy it. Anything that UNESCO can do to satisfy these needs through promoting education, science, and culture will be a step towards a unified way of life and of looking at life, a contribution to a foundation for the unified philosophy we require.

"And, finally, I believe that a body such as UNESCO, which is charged with promoting both the higher activities of man and their practical application, and of doing so on an international scale, is the most likely agency to make this dual approach and so to speed up this necessary process of reconciliation."[1]

In rereading the corresponding document (UNESCO: Its Purposes and Its Philosophy) which Huxley presented to the first General Conference of the nascent organization, the historian Berredo Carneiro says that one feels surprised by the genial intuitions it often contains. He adds that in dealing with questions of education, natural science, human science or just information, Julian Huxley broaches these questions, discusses them, and analyzes them with equal facility and competence. Unfortunately, concludes Berredo Carneiro, the delegations present did not measure up to these intuitions because they did not approve the document. But, although discarded, his views are taking, little by little, a greater part in the projects which today figure in UNESCO's program.

If, upon joining the thought of Julian Huxley and Serge Raynaud de la Ferrière, we had been able to erect two great statues of both, we could symbolize at the foot of those statues an attitude of hope by their respective members about the results of these two men's actions, their thought, and response of the institutions formed by them. One institution, UNESCO, fulfilled its first plans for a headquarters in the Hotel Majestic in Paris. The other, the UGB had the 'grandeur' of beginning in a hut in Maracay, Venezuela. Although the founder of UGB had previously united one group of scientific personalities in France[2], he came to South America to found the first colony of perfection, establishing it in a hut in Maracay, Venezuela. (It was an Ashram, a little in the manner of the Ghandhigram of Ghandhi or the Santiniketan of Tagore, but with its own systems of life, discipline, and investigation.)

The first, UNESCO, was able to count on some of the best human teams selected by member governments, but it would have to make an effort to overcome the limiting bureaucratic walls which would arise in the governments of some countries. These did not allow arousing the collaboration and participation of communities in the extending of their grandiose action. The second, the UGB, was able to count on the direct collaboration and participation of the communities through the force that some economists call intangible and human which arises from the fervor and dynamism of the groups of men in the street, But UGB would have to exert effort to overcome the limitations of its modest and particular classes. At times these classes would be reduced by an individual ostracism which would try to circumscribe the living experience of ontological self-realization itself by refusing to participate and collaborate in the development of high levels of science, education, and culture.

Huxley set forth a governmental UNESCO which would bring a fullness to the future of human communities. Raynaud de la Ferrière set forth a UGB which would be constituted and developed on the highest levels of science, philosophy, art, and didactics, from the individual and community levels, by the learning of synthesis and other matters to perfect human beings—that is, a world united by learning, attained through the union of the sciences and the union of religions.

The first can count on the means, but at times obscures the concrete realization of its objectives. The second appears disposed to its objectives, but at times obscures the concrete realization of the means.

In some of the first members of UNESCO, Huxley found resistance to adopting the philosophy which he proposed because of the religious nature of some governments. Raynaud de la Ferrière indicated in His Circulars, or Letters from France, that his work was finding some members of the UGB on the American continent saturated by religious ideals.

Huxley set forth the scientific basis for a new culture and a new religion, but the scientific investigation itself has not always been sufficiently within the direct reach of UNESCO, especially the response on the part of investigating organizations of a particular character. The UGB, filled with the ideal of a union of sciences (and a union of religions) glimpsed the profound panoramas of scientific investigation, its epistemology, and the transcendence of its educative scope, as well as the re-education needed for a new humanity on the march. It was not, however, able to count sufficiently on the help of governmental organizations and industry. Though both organizations are international in nature and scope, at times they are effectively applied only in a few places—fifteen to twenty countries—in spite of the fact that they have representatives in many more.

Granted that Serge Raynaud de la Ferrière outlined a synthesis for the UGB to make (a real integration of its existence with the proper cultural and scientific program of UNESCO), at the same time he foresaw that the UGB ought to be throughout time an entity of the same level of organization. Its intrinsic nature would also have its existence on the plane of science, education, and culture.

His own literature, as it was catalogued in the *Symposium of Literature* in Guayaquil, Ecuador, is a literature of investigation and of living experience (*vivencia*). His one hundred works belong to a new precessional age; as he said, a new humanity on the march toward a world united by learning.

We could consider his literature as emanating from his illuminated ideas of man's ways, not only in its high level of academic composition and bibliographical reference, but also from his illuminated ontological self-realization which permits the overcoming of limitations. It provides for understanding a new age through different filters. This manner allows his thought to be united and identified in different forms and in different aspects. We are able to do our part in constituting that union, that ontological realization, which is yet subdivided into intervals but always under a logical reasoning.

It should be added that the norms and disciplines freely revealed in his works through the investigation and experience which arises from them, allows this thought to be realized as time advances; that is to say, in the beginning of a new age which is gradually emerging from the stage of transition between two eras of the 20th century. That is, the action outlined and illuminated by his thought at the world level is verified in its prospective.

Though we will consider some other aspects of his primordial thought in the following chapter, let us now turn toward some realizations of the UGB.

Through its first thirty years, the UGB did evolve toward the high level of synthesis with UNESCO which Master Raynaud de la Ferrière sets forth at various times. There is, for example, the progressive establishment of the International University of the UGB which began May 24, 1970 in the form of an open university with free lecture rooms where elective materials about ontological self-realization and interdisciplinary researches were set out. This was directed by a Coordinator Deanship, first in Lima, and then also in Venezuela. The response was favorable in most of the countries throughout the length of Latin America. Soon, in union with INVESCIENCIAS (the Institute for Scientific and Technological Investigations and Applications, established in 1973), with the support of the Minister of Education of Venezuela, we asked SECAB (Executive Secretariate of the Andrés Bello Compact) for educational reforms in the Andean countries which would allow the existence of lecture rooms at higher education levels without professionalistic compromise. This would allow open access to lecture rooms for those who wanted to study university taught subjects for their culturalization without having a university degree as their final end. The Minister, L.M. Peñalver, returned it with a recommendation to the Executive Secretary of the Andrés Bello Compact of the Andean region countries in Bogotá, Colombia, accompanied by the Official seal which signaled the following aspirations and petitions of educational interest for resolving the Fourth Reunion of the Ministers of Education of the Andean countries, and thus awaken similar activities in the other countries of the world:

> "According to the conclusions of the Seminar of Ancient Culture of the Andes (organized by the UGB in Quito, Ecuador in September, 1974), the execution of the Trujillo Compromise, agreed to by the universities in the member countries of the Andrés Bello Compact (Andean Group) on May 10, 1974 in Trujillo, Perú, would be a suitable time for calling a preparatory assembly to incorporate programs investigating the anthropological unity of South America.
>
> "Unifying the Graphs of Quechua and Aymara in the countries of this language facilitates the extension of education, according to the Proposition stated in the 'Seminar on Teaching in the Countries with the Quechua and Aymara Languages". This seminar, with the intervention of the Fourth Region of Education of Perú, was organized by UGB and held in Cuzco, Perú in October 1974.

"The creation and inclusion of Educational Cooperatives at the University and Secondary Education level or its equivalents, with the Laws of Cooperatives, with rights to credit in respective Cooperative Banks, and the creation of a Cooperative Support Fund of the Andean Group for the formation and creation of Educational Cooperatives on their different levels."

In official and emphasized form, the Andrés Bello Compact sent this to Dr. Héctor Troyana, its advisor to the Executive Secretary of the International Symposium, organized by INVESCIENCIAS in Medellín, Colombia, (at which was affirmed the Scientific Act of Quirama)[3]. This Act, on the guidelines for scientific conventions and universities, was in accord with the objectives expressed by the Andrés Bello Compact for Education, Scientific and Cultural Integration of the Countries in the Andean Region.

It was also felt certain that at some time the unedited *Scientific Theory of Cosmobiology: Acts of the Sciences I* was liable to be presented in the course of the symposium by its author, the American epistemologist, Dr. David Ferriz Olivares. The said theory, subscribed to by the Directorate and Vice-Rectorate of the National University of Trujillo, Perú, and by the Latin American Directory of Mathematics located in Montevideo, Uruguay, was submitted to the Presidium for documentary support.

Joined to it were the interdisciplinary scientific goals presented by specialists in branches of medicine, physics, mathematics, astronomy, biochemistry, psychology, and social sciences. Didactically illustrated data and graphs describing cosmobiological art were simultaneously presented to the symposium in the Ricardo Olano and Fernando Botero Rooms of the Zea Museum, with participation by the Department of graphic Arts (University of Antioquia), its Cultural Extension and its Academic Director.

Even though the UGB is international and not locally formed, these significant steps are taken under the imperative, proposed by its founder, that the UGB ought to reach the level of UNESCO.

In other paragraphs Dr. Raynaud de la Ferrière defines some aspects of the UGB as founded by him:

"Since 1948, the foundation of Dr. S.R. de la Ferrière, an international institution now known for its benevolent social acts, has been manifesting itself in all aspects of life: assistance to the poor, help for the disinherited, the setting up of dispensaries, libraries, schools for children as well as for adults, sporting clubs and art centers. It cooperates also with other organizations such as UNESCO, the Red Cross, Scouts, and other humanitarian associations.

"Presently represented in fifty countries, UGB extends its activities to five continents through conferences, study centers, journals, and diverse works. It is a cultural movement based on the theme: SCIENCE-PHILOSOPHY-ART-DIDACTICS, and is presented as the vanguard of a New Age which began to reveal itself with the re-education of humanity through the synthesis of knowledge.

"Millions of members are already working for this great endeavor of the UGB which pursues as its goal the science of living well, the art of existing correctly by developing a program for the closer association of races, peoples, and creeds. In this pursuit of a common ideal, it seems to all to be preparing mankind for a better destiny".[4]

"In these moments when events seem to exceed the limit of current sociology, when politics and religions dispute the supremacy of peoples, it is a duty of the Assemblies of learned men to take direction of a world organization for revealing the truth.

"Since the ancient times of man, a community of beings has evolved for giving men a knowledge of their true mission.

"In this New Age, humanity finds, at last, a complete movement that is not going against any creed, but which symbolizes the aspirations of everyone in a synthesis of concepts.

"The institution does not pretend to establish a world government, but wants to study all facets. Consequentially, a new understanding and cooperation in the entire human community would be seen with a very great degree of pleasure.

"The UGB conceives, above all, a world ideal uniting all nations under the sovereign direction and council of an Assembly of Wise Men, who are highly talented intellectually and spiritually. These men, with a greater sense of justice, would study, always on the world level, the correct distribution of services, the natural riches of the earth, and the products of technical and social economies for both assisting the general community and for particular elevation in the domain of the individual.

"With the purpose of studying and determining what is possible in solving the problems of education, scientific research, and other aspects of culture in general, the Universal Great Brotherhood was entered in the Register of UNESCO (Section II, Category 3) in Paris on September 15, 1954, and afterwards on September 23, 1958, another certificate was extended to it.

"The institution today is represented in a large number of countries on the five continents and is administered on the one hand with the help of a Supreme Council (assisted by a Judicial Council) and organized, on the other hand, with the help of Executive Councils. These Executive Councils direct the sectors found under their jurisdiction, always informing the appropriate authorities in their zone, city, and country. They always attempt to conform to the current national laws, under the support of a judicial solicitorship or legal registry.

"In each country, instructors have been greeted with as much approval by some official persons (chiefs of state, ministers, directors of educational services, presidents of cultural groups, etc.) as by some religious authorities (prelates of churches, respresentatives of sects, superiors of spiritual communities, etc.). To all they have come to join the men of science with the humble, animated, people in achieving the happy ideal of human brotherhood.

"They have effectively organized international conventions and count four already in America: the first in San Salvador in 1954; the second in Tegucigalpa, Honduras in 1955; the third in Mexico, D.F. in 1956; and the fourth in Caracas, Venezuela in 1957. These congresses proved the proposition of establishing bases for a new humanity can be successful. As a result, some persons have offered their valuable assistance in the urgent work of human re-education in order to achieve a better world, of understanding and harmony.

"Various countries have offered land to the institution for the creation of colonies where artists can work in a favorable atmosphere, scientists can pursue their research with the necessary tranquility of spirit, and mystics can speak and meditate in peace about the correct ideals of good—all separated from noise and other environmental disturbances.

"Some of these colonies, often called ashrams, have extensive areas separated from the crowded cities. There are gardens specially cultivated for medicinal plants, fields for sports, gymnasiums, athletics, and other healthful recreations. They also organize schools for literacy, infant care centers for treating and helping the mentally retarded, orphans, and abandoned children; dispensaries and other centers for naturalistic treatments. There are rooms for showing educational films with the assistance of competent authorities. Some films are loaned by consulates, embassies, UNO and petroleum companies. They are always accompanied by instructive moral conversations. They also produce cultural, theatrical, and cinematographic activities for procuring funds to help the beneficial branches.

"These 'Ashrams' (or colonies for spiritual advancement) are always built according to well studied plans, and installed with natural comfort which permits those who live there to be always in intimate contact with Nature; to work and live in the sun, near rivers, and forests, and other pleasant places.

"These 'Ashrams' are numerous in the Orient, especially in India, and are becoming increasingly common in Europe and America. Generally they are co-educational and education in the more perfect forms of ascending spiritually, intellectually, psychically, and artistically—is offered to the children. The number of residents ranges from a dozen to five hundred, although generally they do not accept so many when they deal with children. These centers of perfection have been accepted very satisfactorily by various sectors and social organizations, including some governments, that have given facilities and economic help to such colonies.

"It is for those in need of renovation and dynamism in their search, of documentation for their elevation, a direction for their evolution, care and counsels for their physical and moral health, etc.

"Some sections of the UGB operate from noon until midnight. By means of a permanent secretary, they regularly offer information to the public, along with a library, care for the sick, reunions, courses, and conferences according to the degree of development of the attendants.

"Officially functioning in more than thirty countries, the UGB pursues its activities on five continents, always trying to strengthen its own centers while at the same time expanding its radial operation with new centers of study.

"Some countries work with various Subordinate Councils under the direction of an Executive Council, which represents the Supreme Council. Other regions (especially Europe and Africa,) are only represented by small groups of sympathizers.

"It would be ideal, and is almost indispensable, that the Supreme Council of the UGB take the necessary steps to help those poor small groups.

"It ought to organize more public activities, such as recreational festivals, and theatrical functions. With the assistance of artists who are sympathetic to the Mission. The purpose of acquiring funds for developing the social services of the institution could be served.[5]

"There are other activities which also are part of the cultural centers of the Institution.[6]

"Some Presidents of Republics have not hesitated to intervene in support of our organization. In the United States, some artists gratuitously have given their assistance for festivals; in Venezuela, various ministerial departments have offered help for the establishment of a colony. In Burma, the proprietor of a maternity hospital donated his establishment to the UGB. In Australia, a medical person donated a piece of land and offered his clinic free to the Institution.

"Although each sector has its worries, its needs, and its obstacles, yet from everywhere persons of good will come to lend support according to their own capacities. Proprietors offer some goods, lands, premises, villas, furniture, offices, etc.; others put funds at the disposal of the work and some offer their labor and abilities. Editors and printers propose to publish for free the books of the institution; others freely translate the works. Attorneys and doctors offer their services without expecting fees. Generous offerings are received in some parts, but the UGB needs to be strengthened still more in this kind of spontaneous collaboration.

"Others can offer their wisdom, their collaboration in a special work; for example the institution needs secretaries, lecturers, professors, workers, farmers, etc. this is the part which pertains to man in his mission of universal brotherhood.

"Nothing ought to be ignored, now that we have entered a New Age. If some feel their evolutionary influence in an unconscious manner, others have become fully aware of it as a psychological phenomenon. Among others, men of thought have felt 'commissioned', and after detaching themselves from the great human mass, they become archetypes symbolizing the Great Civilization which they already glimpse.

"On the other hand, one appreciates more and more the desire to fuse, to fraternize, to unify. The great movements make concessions, but organizations of all kinds have to achieve results. Clearly the Universal Great Brotherhood expresses the general ideal of men and manifests this tendency more frequently each day.

"The Institution throws out again its 'call' to all organizations that want to collaborate in the same great work and common ideal.

"It is time to end the strife which divides and squanders the energies, when such organizations work for an identical cause; it is time to join forces for sooner achieving a task seen and desired by all for the well-being of humanity.

"The inspectors of the UGB, like missionaries, are at the disposal of groups for instructing them on the science of life. For many years some of these instructors have been offering their services, going from country to country, setting up public conferences on diverse matters, as often in colleges and official institutions as private groups. These active persons of the institution, besides feeling predisposed for the missionary life, were educated in the centers of the UGB. After a residency in the ashram under the close direction of the Master, they have been consecrated with total abnegation to impersonal and constant service in the great cause of human brotherhood.

"The members of the UGB do not comply because of any obligation, but because of a purely moral beneficence, thus remaining still at liberty.

"Yet the present founder of the UGB, Dr. Serge Raynaud de la Ferrière, does not ask that one follow his example.

"It is natural that in the habitual development of deeds, there is always a necessity for the presence of a leader who generally serves as an emblem or symbol to the followers.

"Finally, the great organization is involved in developing the perfection of the human race, with the supreme hope of a better world in an era of tolerance, truth, and peace."[7]

In the Great Messages, he set forth in detail:

"All those who speak on behalf of the Universal Great Brotherhood are easily recognizable: they study what their neighbors think, and are interested as well in all the more beautiful examples of tolerance. To believe that 'one's' religion is the only valid one is pretentious; the 750 million who follow Buddhism and Brahmanism have never thought of imposing their ideas on the rest of the world. On our planet, among the number of inhabitants, who vary between 2,000 and 2,500 million (in the decade of 1950), are a great quantity of people who are Mohammedans, Jews, and Christians. To ask the suppression of religious borders is to be inscribed in the UGB by right itself.

"The institution does not ask anyone to adhere to a special doctrine because it does not act like those who criticize primary school when they pass to the secondary courses. We would never deny the interest needed to learn the alphabet, although we are already University students.

"One does not try, then, to abandon 'his' religion in order to be a member of this great institution; one tries simply to be elevated in spite of all the human contingencies. "It is necessary to come to the point which men on the level of the Great Principles understand, as Claude Bernard said: ". . . when the scientist, the historian, and the philosopher speak the same language . . ."

"Dispersion was the origin of the religious dissension. The differences in language brought about diverse terminologies for the same thing. Analysis reveals the special character of each group of study and all, naturally, are invested in religious doctrine which is the basis of all the barriers among men.

"To return to the setting forth of religious unification, it is certainly difficult when one understands it according to the letter, but when one takes it in accordance with the spirit, one can admit that in spite of an appearance of contrary points of view, all come from the same fount."[8]

Let us turn to the comparative considerations between both wise men. What would they expect from the colossal symbolic statues, which we previously erected in the middle of the garden of a human city, as an expression of the attitude of their institutions members?

If we imagined their statues as separate, we could suppose that the respective members of each Institution would be strengthened by considering the opposite figure. The follower of UNESCO would be turned toward Dr. de la Ferrière for achieving his plans to reunite the dispersed truth and join it to the living contribution in each field of human universality, by means of his 100 books wherein is found a synthesis of ancient and contemporary knowledge for awakening the individual conscience toward union with existence and the universal conscience. At the same time the members of the UGB would be directed toward the statue of Huxley because the UGB fosters the organization and achievement of the highest level of science, education, and culture a little in the manner in which its sister, UNESCO, did in its first 30 years.

But if in imagining the statues separated, we then imagine them joined we would find:

The answer to the call of Master Raynaud de la Ferrière, for a synthesis of the work of UGB and UNESCO, comes through the systems and criteria of the highest level for achieving a world united by learning.

In Huxley, Paulo E. de Berredo Carneiro says, the man of science, the philosopher, the apostle, came to occupy little by little, a place which belongs to the rare personalities of our time. And let us not doubt, he adds, the admiration and gratitude which posterity will show him by subscribing to the judgment which Huxley made of himself: "If I have left a memory of myself, I hope that it will not be fundamentally as one specialized in scientific work, but as a 'generalist', someone to whom paraphrasing and expanding the phrase of Terence, 'nothing which is human, and nothing which exists in Nature is foreign'."

NOTES

1. Julian Huxley, UNESCO, Its Purposes and Its Philosophy, and A Philosophy for UNESCO; The UNESCO Courier, March 1976. (Vol. XXIX) pp. 15, 16, 18, 23, 30, 33-5.

2. The group of scientific personalities in France, the group which was united for a first contact with the public, was:

Professor Toutain, Director of the University of Sorbonne, Paris
Professor de la Tramereye, President of the European Polytechnic Society

Abbot Moreaux, Astronomer, Director of the Observatory of Burgos
Professor Prunier, Engineer, Chief of Technical Service of the French Railroad
Professor Boutaric, Mathematician of the Faculty of Sciences, Paris
Rev. Father Dopp, of the University of Louvain in Belgium
General Ducroc, of the French Army Air Force
Professor Varcollier, President of the Physics Society of France
Mrs. M.E. Cluzel of the Geographical Society of New York
Leo Verhardt, Technician, and ex-Counselor of President Roosevelt
Dr. Sauvageot, Medico of Paris
Professor Bourges, of the Faculty of Algiers, North Africa
Dr. Encausse, Inspector of Health of the Parisian Colleges
Colonel Foss, friend of President Eisenhower
President Herriot, who was Chief of the French Cabinet, President of the General Assembly
Consul Ellessilinger, of the United States in Paris
Professor Federico Emilio Tranin, of the Charcot Group, Polar Explorer
Professor Marchand, of the University of Mexico
Marcelo l'Herbier, film producer in France
Professor Ananoff, President of the European Astronautical Society
Mr. Lavachery, Director of the Colonial Museum of Brussels
Minister Monerville of the French Government
Federico Dupont, Deputy to the Government of France
Professor Kamiensky, of the Observatory of Krakow
Royal Segun, Consul of France in Sweden
President Poncelet of the Government of France
Professor Sommerfieldt, Danish explorer and writer
Dr. Angel Baldizon, of Guatemala
Professor Claudio Urrutia, Director of the Observatory of Guatemala
Mr. Zadelhoff, of Holland
Count H. de Souza, of Portugal
Prince de Broglie, of Marrakesh
Professor Zavagno, of Italy

3. Signers of the Scientific Act of Quirama:

Dr. Héctor Troyano Guzmán, Representative advisor of SECAB, the Executive Secretariate of the Andrés Bello Compact
Liliam Palacio de Alzate, President of the Departmental Assembly of the Department of Risaralda, Municipal Council. (Colombia)
Dr. Carlos Alberto Irígoyen, Peruvian Diplomat, Representative

of the Faculty of Cosmobiology, President of the Executive Council Perú

Dr. Jorge Rodríguez Arbaláez, Director of the Institute of Cultural Integration

Carlos Corcull Boada, Representative of the Supreme Council of the UGB, Headquarters in Caracas, Venezuela

Representative of the National University of Colombia (signature illegible)

Rector of the University of Medellín (E), (signature illegible)

Representative of the University of Antioquia (signature illegible)

Mario Flóres, Engineer, Representative of the Higher Institute of Cosmobiological Investigations of the UGB in Buenos Aires, Argentina

Dr. Eduardo Gómez, Representative of INVESCIENCIAS of the UGB, in Medellín, Colombia

Professor Orlando Rodríguez, Representative of INVESCIENCIAS of the UGB in Medellín, Colombia

Professor Consuelo Hernández de P., representative of INVESCIENCIAS of the UGB in Medellín, Colombia

Miss Nubia Gaviria, Representative of INVESCIENCIAS of the UGB in Medellín, Colombia

Elmer Trujillo, Engineer, Representative of INVESCIENCIAS of the UGB in Lima, Perú

Mr. Fernando Florez, Representative of INVESCIENCIAS of the UGB in Caracas, Venezuela

Dr. Graciliano Arcila, Representative of the Anthropological Museum of the University of Antioquia

Professor Edwin Rodríguez Paz, Representative of the Executive Council of Colombia

Javes Vallejo Valle, Engineer, Representative of the Subordinate Council of the UGB in Medellín, Colombia

Dr. Wendel Polich, Honorary President of the First International Symposium of Cosmobiology

Dr. David Ferriz Olivares, President of the First International Symposium of Cosmobiology, Director of INVESCIENCIAS, (Institute of Scientific and Technological Investigations and Applications), Sub-Director General of the UGB

4. His Instructional Letters, Tome II. p. 6.

5. A New Humanity on the March, pp. 5, 9, 14, 17-23.

6. In the Artistic, besides founding Art departments, poetical recitations, concerts, presentation of folklorical assemblies, staging theatrical works, expositions of art; painting, sculpture, and ceramics, editions of periodicals, reviews, wall newspapers, organization of libraries, discotheques, artistic collections. Performance of cultural acts, group

discussions, holding of round tables for discussing problems of current value, organization of choirs and glee clubs.

In the Scientific, besides scientific research, organization of small museums, expositions of scientific character, making up of herbariums, terrariums, aquariums, planetariums; organization of outings, walks, and excursions with the character of study and recreation, for which it will have an excursionist organizing director. It also gives increase to other activities which stimulate the aptitudes that accredited members of the institution have for scientific work, and that yield an effective contribution to the research, understanding, and divulgence of sciences, conferences, and educative articles.

In the Social, organization of professional courses, organization and maintenance of primary schools, for children and adults, day and night, secondary high schools and institutes of higher and technical education, cultural centers, foundation of consumer cooperatives, savings and loan banks; first aid chests and other special services, such as: medical-dental consultation, legal conferences, etc., making use of the professions of such members and sympathizers of the institution. Collaboration with all those campaigns which aspire to regain the community; collaboration with the campaigns of public good, like those of hygiene and anti-alcoholism, protection to the animals and plants.

In the Sporting, organization of teams, Baseball, Football, Volleyball, Basketball, Tennis, Ping-Pong, etc., organization of championships. Athletics, cycling. Conditioning of playgrounds and Gymnasiums; donation of trophies. Practice of 'Bola-criolla' Organization of parlor games; chess, dominoes, chinese checkers, courses of judo and jiu-jitsu. Practice of physical culture and edition of gymnastic reviews.

In the Festive, performance of dances and folklorical festivals, evening festivals, picnics, lunches of entertainment; rural walks, excursions to the beaches; visits to other fraternal institutions. All kinds of celebrations, from the solemn extraordinary session of commemoration to childrens parties.

7. A new Humanity on the March, pp. 27-32.
8. The Great Messages. pp. 333, 362-3.

CHAPTER 5

Does Popular Resistance Disappear at the High Levels of Science, Education and Culture?

Environment, living experience, and creativity. They intertwine somewhat as Marshall MacLuhan states, referring to the fact that the smallest bit of technology forms a new environment. For example, the earliest appearance of the chair was the cause from which the table resulted which in turn brought about a new arrangement of the people seated at it. Or, as happened with the wheel, which gave rise immediately to another technological form—the road, which also changed social standards.

All code and mathematical language have magical properties which transform and transfigure; for the electronic age is inclusive—it includes everything.

The resurgence of mathematics presents us with a new factor, namely science, which impels our society to abolish the abyss which exists between the man of science dedicated to his own specialty and the general public who, for lack of explanations, lives plunged in ignorance and looks toward science full of superstitious fear. At the same time, they ask it to produce miraculous objects and remedies that a certain kind of popular press has taught them to expect from science; and not only to expect, but to consider as the inevitable result of their investigations.

Without a healthy understanding of the methods and procedures of science, the apparently fortuitous character of the discoveries that science makes stimulates the public to adopt an unhealthy attitude toward it, an attitude full of mistrust. They always fear what they don't understand.

If the man of science needs long and detailed studies for learning, how can he expect that we immediately understand what he says?[1] Jan Comenius, moving into the New Age, after 1640 had proposed that science tie itself to learning. For this reason he considered it necessary to create a 'pansophicon' where the sages of the entire world would meet for a year to present, evaluate, and examine the facts about nature gathered from throughout the world, for the sake of exhibiting and publicizing them, which would permit their utilization and learning by men in general.

This is something which could have been inspired by the Yachay Wasi (House of Knowledge) of the Incan Empire or by the House of Solomon which Sir Francis Bacon spoke of in his *New Atlantis*, and which would now require the fundamental participation of the scientific press. It is like the need of a heart to pump the elements of science throughout the whole social body and to induce the spirit of curiosity and a universal diffusion of knowledge.

Fortunately, one already observes that even illiterate peoples appear eager to learn. This gives rise to a progressive elimination of the obstacles opposing the progress of the popular scientific environment, one of which has been public ignorance.

Nevertheless, it is not surprising that the uninformed consider science as a kind of bank vault to which only those who possess a university degree in the subject have access.

All the great savants have known how to make themselves understood. Even Einstein exerted himself to achieve it. The second-rate men of science are those who do not completely dominate their subject and who become unintelligible because of a defensive reflex.[2]

One also notices the collective fact that a large number of adults, as children, suffered traumas or psychic blows from authoritarian primary or secondary teachers, above all, those who had to forsake their studies (especially those who later found themselves dependent to some extent on people of better education.) A submerged hostility easily rises to the surface when relations or attitudes appear toward some professionals who, justly or unjustly, are the object of psychological transference of such psychic contents. Men of knowledge have suffered more than once from the attacks of those who have a lower cultural level, in addition, to a lack of understanding, and even alienation at times. They are like an anamorphosis, or deformed image drawn on a plane surface, which only shows from one point of view an image according to normal vision. Human behavior is a reaction to the images that each person forms in his mind and produces in relation with a specific environment, whose representation is not the same for two different individuals. It is said that cultural circumstances interpose themselves between man and true external reality.

The problems which arise due to different cultural levels are harder to solve because they depend on many factors which flow together like the tributaries of a river: our education by means of different words, our transpositions of real life, the different meanings of behavior, the feelings of displacement, and the rapid evolution of the New Age.

One of the sharpest aspects of this transition between two precessional eras, which includes the 20th century, is the incompatibility between the different cultural levels and their disregard among themselves. If in previous times, a wise man like Galileo was persecuted because his discoveries were contrary to the dogmas, or Ptolemy was scorned in front of Copernicus through the complete disregard of the topocentric mathematical models of the modern advance, the dawn of the new age shows the separativity between the different levels of cultural profundity. This is shown not only in terms of lack of understanding and rejection by the great religious sects in the presence of figures like Teilhard de Chardin, but in the governmental sectors which, above all in the first half of the 20th century, minimized the revolution of the intellect.

The highest cultural level has been confined to elitism through the pressure of the other levels and as its own answer and defense of the most profound level of studies.

He who does not know, does not know that he does not know, and has either disbelieved, or has supposed contrary or such favorable fantasies, that he verges on the apologetic about what he knows so that the psychology of rumor has not favored the true approach, the understanding. The power of ignorance delays the decisions and advances of an age of knowing and on the other hand, the man of true investigations has taken refuge in the apostolate of his disadvantageous work which is yet pitifully like a calvary which he lives in the redemption of knowledge.

The UNESCO director of the International Institute for the Unification of Education, Phillip H. Coombs (former Secretary of State for Education and Culture in the United States during the presidency of John F. Kennedy) indicated in *The World Crisis of Education* that an excessively large number of the brightest European students prefer a general and humanistic education on the secondary and higher levels, turning away from the professions whose fundamental studies are the sciences and mathematics. This attitude can be blamed on their natural aptitudes or on their having had incompetent teachers teaching these subjects, teachers who failed in their task of giving encouragement to their students. It indicates inclusively an annoying phenomenon that good students appear to form a group opposed to mathematical and scientific studies; in the face of this, educational authorities have encountered difficulty channeling a greater percentage of students toward scientific and technological studies in order to offset the imbalance. On the other hand, an educational system and its students cannot be asked to do what society itself is not capable of doing; namely, encouraging the proclaimed objectives. But one can, at least, ask education that it not contribute to fostering attitudes and preferences contrary to the vanguard of development.

Education, naturally, cannot by itself assume the whole job of reforming society in its attitudes and uncompensating structures because education itself, being conditioned by the same society, ends up expressing the same thing.

On the other hand, in various underdeveloped countries there exists a strong tendency on the part of students to prefer bureaucratic jobs or jobs 'behind a desk' because manual labor, since it is in the field or in the factory, is associated with a lower social condition. The schools and the teachers tend to reinforce this attitude instead of trying to modify it.

Even when they try to reorient the educational system, the efforts of daring teachers can be nullified by opposition from stronger persons who are deeply rooted in their respective societies or institutions.

When the abyss between the man of science and the general public is abolished along with the levels representing this in the governmental and private sectors and when a true contact is established between the cultural elite and the ignorant masses based on the concepts of the new age, the Age of Knowledge will have begun. The real thought and illumination of men will be applied at every level and expressed by, for example, the Master Raynaud de la Ferrière.

By the way, when he answered one of my letters in which I asked him about the possibility of adapting some deep subjects from his texts to those who had only incomplete primary studies, he expressed to me that he very much regretted that many of his teachings required at least a secondary education to be perfectly understood.

Apropos of this socio-educative imperative and to understand more clearly the framework in which we are presenting his thought within the short space of this book, it is necessary to explain some aspects of the ideation of Dr. Raynaud de la Ferrière regarding the levels to attain in science, education, and culture.

Let's try again to understand the thought of Dr. Raynaud de la Ferrière with regard to the mathematical environment of the present and in the past. We shall see then the necessity of making references and descriptions in his own expressions that mention the socio-mathematical structure which surrounds it, that is, the fount, in his depth and original writing for surrounding it in the framework of his living experience (vivencia). Thus we secure in both cases the suppressing of commentary on the experiential panoramas in his own expression, that is, the words of the Master and those of his age: the past, present, and future.

NOTES

1. The UNESCO Courier, February, 1965. p. 5.
2. Ibid. p. 8.

PART II

THE MATHEMATICAL ENVIRONMENT

CHAPTER 6

The Mathematical Expression of the Ancient and Modern Social Structure

In summary, let us analyze the conditions of the socio-mathematical framework. David M. Messick says that during the past twenty-five years persons working in the social and behavioral sciences have seen the necessity of depending more and more on mathematics as a conceptual tool for facilitating the understanding of peculiar phenomena in each one of these disciplines. The number of fruitful applications of mathematics to such sciences is increasing rapidly.

In the first instance, let us consider the subjective content, relative to mathematics, which is brought to the living experience (*vivencia*) from something abstract, or from a concept of divinity, or from a horizon of creativity.

First, we must consider its repercussions not only in the interior, subjective life, but as a factor of the environment, an experience which comes from without and which sociologically influences the medium, transforms society, and consequently affects the individual, who is not exempt from that influence, however indirect.

In the case of the person who has a living experience through mathematics, a direct influence results, and therefore we observe its effects on ethics, and certainly on the behavioral sciences as well as on the behavior itself, just as in necessities and desires.

This brings us to a study not only of some recent discoveries, but of the conditioning environment which penetrates the man of this new age. This will greatly elevate the standard of that experience and characterize it by one of the phases of future man for whom mathematics can also offer a series of important living experiences, which is one of the purposes of this book.

John Cohen, Professor of Psychology at the University of Manchester, who has a mathematical orientation, points out that they have investigated, by experimental methods, the subtle processes of subjective probability in which the idea of 'I demonstrate' is an essential element at the moment of making concrete decisions, and can include the basis of thought itself. We use our mental antennae to explore and touch the universe; and from these signs, which provide us no more than partial information, we learn to make solid judgments.

"Uncertainty", says Cohen, "fills our lives to such a degree that it dominates our language. Everyday vocabulary is full of terms such as 'probably', 'may', 'soon', 'large', 'small'. How should we interpret such a word as 'probably'?"

To provide an example, nowhere are the processes of subjective probability more clearly shown than in gambling. It deals, of course, with one of humanity's most ancient and universal practices and appears to be allied to the ideas of fate as well as primitive justice and prediction, along with man's reverential fear of unpredictable events.

This study, accompanied by cosmobiology, as a cosmic biocybernetics, gives us a new means of studying mental development from childhood and offers a simple conceptual scheme for studying our methods of perceiving, thinking, deciding, and acting. Likewise, it offers new possibilities for comparative studies about abnormalities in thought and behavior.

Cohen adds that, "It seems important, during the school years, to teach children to distinguish degrees of uncertainty, and to teach them to compare their private predictions and extrapolations with what really happens; in a word, to teach them to be masters of their own incertitude.[1]

Dr. Raynaud de la Ferrière expresses an advanced concept of education when he says:

> "The important thing is not just to teach something to a child, but to mold his spirit for observation and reflection, criticism in research, and the love of truth. 'There is nothing more beautiful than the truth', Boileau has

said. Let us form spirits appropriate to the synthesis. The New Age which will be outstanding from its research of synthesis, will need broad spirits, strong, and capable of understanding a system as a whole. And this possibility depends a lot on the character formation during childhood."[2]

In a significant manner, digital computers are another source of the living experience which we are studying and which ranges from ancient systems to the modern technology of probabilities. These computers are revolutionizing science, technology, and society itself.

Computers are mathematical machines, i.e. machines capable of doing mathematics. They have stimulated mathematicians to think very much about mathematics as suitable for computers.[3]

John McCarthy writes a paragraph of prophetic value for a not-too-distant future in his article "Information", saying that:

"As has happened first with black and white, and later with color, television, it can be supposed that the first ones to install computer equipment in their homes will be enthusiasts and the rich. Nevertheless, all the world will finally consider such electric servants a necessity. People will soon grow tired of disposable, 'prepared' programs and will want to write their own programs. The knowledge to write a computer program will be as common as driving a car."

"Not knowing how to program will be like living in a house full of servants, without knowing how to speak their language. Each one of the programs prepared and presented to the user will be a tool by itself. Nevertheless, it will be up to the individual to combine them for his own maximum benefit. People will find, in practice, that being in front of a control console will lead directly to the writing of their own programs.

"As a skill, computer programming is more difficult than driving a car, but easier than flying a plane. It is harder than arithmetic, but easier than writing correct English. It doesn't require long studies. Many people learn how to write simple programs after one or two hours of study. It will pave the way for obtaining quick success, and this first success strengthens the spirit. It is much easier to learn computer programming than to learn a foreign language or algebra, for example.

"Success in writing a program for doing a special task, depends more on understanding the work itself than mastery of the technique of programming. Planning the route of a comet, for example, requires several weeks of study in programming and several years of studying physics. Writing a program to accomplish a task means the individual involved must explain just what is wanted. The computer for public use, will do exactly what it is asked to do within the limits imposed for protecting the interests of others."[4]

At the same time, Herbert Marcuse speaks of the Great Rejection, the revolt of contemporary sensitivities against industrial techniques and the over-repressive society. But Francois Perroux exclaims, in reply to Herbert Marcuse:

"Science and poetry were never realms without communication. Rarely in our universe, artificial and fecund in material or mental devices, has it been more decisive for the service of men and for the master works of this art than today when the poet knows to pour out his poetry in equations.

"In this eminent society, the social relation is not only a bearer of over-repression and alienation in the technical life, it also bears witness to the secular conquests of the humanization of life. The traditional city not only teaches how to oppress, to exploit, to kill, or to abandon. It also teaches—imperfectly, incompletely, but beautifully—the art of recognizing others, of paying attention to one's existence, both exterior and intimate, of the attempt to hold dialogues that reduce the inequality of dialogue, of respect for experience, and of procedures examined by reason.

"If the disciples of Herbert Marcuse wish to give a universal scope to their esthetic assertion, they should use these indescribable powers of reason which require rigor and mental discipline, from which it does not seem that an alliance can begin with the shameless Great Refusal.

"In the long history of humanity, prophets and saviors, who have brought some peace and sweetness to all or spread nostalgia for these blissful ones, certainly had a very different appearance from these."[5]

These last words of Francois Perroux, economist of the College of France, make us recognize the need for a complete vision which encompasses the ancient and the modern. This includes mankind's contributions from antiquity where the golden mean is in the geometric progression of the ideal archetype of dynamic growth whose reflections, projections, and materials come to be regarded as a harmonious relationship of analogies and calculations. Now, modern mathematics permits biologists to analyze and follow the multiplicity of signals emitted in the interior of the head of man which is a space not very large, and strictly limited, that has to function simultaneously with a number of operations so much greater than that which can appear in whatever other system of the same size and truly great complexity. It employs codes which make its analysis go well beyond the standard of the methods of other epochs, since it is not able to study a unique variable isolated from the rest. The living brain is not able to brake its activity, nor can it forget its past, whatever stimulus acts on a nervous system changes it permanently, but it also unites past, present, and future.

Yes, the past, the wisdom of the ancients. As Paul Valery said, in his beautifully precise letter to the architect Matila C. Ghyka, of the need of comparing natural structures and architectures and the constructions of artists, the apparently arbitrary mathematics that appear in the first formulas, which serve in the arts. And as the same Ghyka points out fundamentally in the Golden Number, I. (The Rhythms, continuation and extension of his Esthetics of the Proportions in Nature and in the Arts): It is the inclusion of mathematical theory of the form, the ideas of proportion and harmony, and the geometric canons which in the great ages of Mediterranean art have been able to serve for the composition of the architectonic outlines, the Pythagorean Theory of the harmony of the spheres, the philosophical and cosmogonic ideas enunciated by Plato in the most Pythagorean of his dialogues, the Timaeus, and with the speculations derived from the analogy, the correspondence between the Macrocosm and man, the Microcosm.

The mathematical meditations of Plato are not extravagances which can be isolated from his doctrines but the living fount of his conception.

Plato has a conception of number when he says in the *Epinomis* that the numbers are the highest grade of knowledge itself, and that everything is disposed in accordance with number, starting from Pythagorean thought. The mathematical theory of today begins to look like that of Plato, of Nicomachus of Geresa, and the Sacred Discourse of Pythagoras.

As pointed out so well by Matila C. Ghyka in his interesting work *The Golden Number*, Nicomachus, like Plato, 'distinguishes two classes of numbers: the Divine Number, or Idea of Number, and the scientific number',

> "The first is, naturally, the ideal model of the second; that is to say, of that which we consider generally as number; but the cause of what in the material world are the forms (which depend on quantities, qualities, and dispositions) the only permanent things, and of which the structure of things (a copy of the model or paradigm perceived by the Logos as resulting from the Idea and from Number), is its unique reality. It (the Divine Number) will always be, generally speaking, the Archetype director of all of the created Universe."

We observe first that the same word Logos signifies in Greek, reason, reasoning, and relation. (The judgment, essential faculty of the reasoning intelligence, is, for the rest, the exact perception of the relations between the ideas and things.) This same term, the word for essence (as later the term 'Word' in the Fourth Gospel), signifies also the divine creative intelligence. Nicomachus will indifferently identify the creative God with the following: that which composes with art, that which makes simple recomposition from chaos, that which selects.

Now it is synthesized by the concept: Divine Number, or Ideal Number as scientific number; they are not yet separated, but are the union of the investigation and the living experience (vivencia).

In the *Theologumena Arithmeticas*, Nicomachus treats of this Number-Idea, or Pure Number, and in the *Introduction to Arithmetic* of the scientific number. And thus, the theory of numbers was divided into two disciplines: the first, Arithmology (Mysticism of Number) of metaphysical tendencies, which concerns Pure Number; and the second a Theory of Scientific Numbers which treats of abstract scientific number, according to a rigorous Euclidian type syllogistic method which is directed at the philosopher, not the beginner. Finally, he had a third science, or better, a technique relegated to an inferior grade, (which today we call Arithmetic) which was the Calculus that dealt with concrete numbers.[6] A scholium on the Charmides of Plato says with respect to this:

> "Logistics (calculus) is the theory which is concerned with the numerable objects and in no case with the (true) numbers. It does not consider, in fact, number in the proper sense of the word, but supposes that 1 is unity, and that all which can be enumerated is number (so in place of the triad, take 3, in place of the decad, 10) and thus applies the theorems of arithmetic."

This distinction appears much clearer if we remember that the Greeks did not employ exclusive symbols, or cyphers, for representing numbers, although they were concrete. The letters of the alphabet were used, and some supplementary signs (the Pythagoreans employed in Sicily groups of points, which they brought directly to the stereometric properties of numbers and to the 'figured numbers'). Arab cyphers and the decimal system made calculation (what we call arithmetic) so easy that we forget the distinction between philosophy of number, theory of numbers, and calculus, as well as the difference between cardinal and ordinal numbers. We have to thank Cantor and Russell for the creation of the theory of sets and for discovering anew that the cypher 2, the number two, the dyad or pair, and the idea of duality are very different things. Let us then, forget the cyphers and thinking about pure numbers, and it will then appear as reasonable to us as to our two ancient guides that, the Cosmos is ordered and rhythmed, and number is, according to the expression of Nicomachus, the eternal essence of reality.

The principles (origins) of number and of the rest of all things, are he says, copying in this also the terminology of the Timaeus, 'the same', and 'the other', (or quality of 'being the same thing', or of being 'another thing').

Moderato of Cadiz (a celebrated Pythagorean and mathematician of the age of Nero, and author of a lost treatise titled *Pythagorean Scholios*, cited by Porphyry) says, "The Pythagoreans called 'One' the idea of identity, of unity, of equality, of concord and of sympathy in the world and 'Two' the idea of 'the other', discrimination, inequality."

Modern science has just come to a spiritual attitude analogous to suppressing again the barriers between mathematics and logic: the Theory of sets, of classes and relations of Cantor-Russell-Whitehead, and the Axiomatica of Hilbert, are chapters of a unique science. They are the new "Logistics", whose elements and symbolic characters represent indifferently logical fictions, and numbers of geometrical configurations.[7]

Nicomachus of Geresa thought that all that nature had arranged systematically in the Universe seemed to have been as arranged in its parts as in the determinate whole, and put in order to accord with number. And afterwards he adds in his *Introduction to Arithmetic* that it was done by the prevision and thought of That One who created all things and who considered the model as fixed, as a preliminary outline for the domination by the pre-existent number in the Number of the world's Creator God. This number-idea, the true and eternal essence, is manifested in accord with number and time.

With respect to scientific number, Nicomachus said that it is a limited multitude: a combination of monads, unities, magnitude, and a multitude of different elements like quantity, quantum, and the volume which flows.

Theon of Smyrna was an ideologue who was dedicated to facilitating the study of Plato through his writings, and what he contributed to Platonic thought was known and valued in his time and for posterity, as it happens with each great master.

Theon of Smyrna relates that the doctrine of Pythagoras was condensed after his death in the Doric verses of Ieros

Logos whose boxes of greater hierarchy were formed by the adept mathematicians. But he also had cultured followers who were recruited democratically, but who gave Pythagorism a fanatical and puritanical twist of practices which seemed to have degenerated into puerile superstitions, and these were called acusmatics, including some who mocked the comic poets.

The mathematical Pythagoreans, according to Theon, considered the natural series of numbers as the principles.

Matila Ghyka comments that it is almost literally the definition of number of Bertrand Russell as 'class of classes'. Also he says that the number-cut, or number-segment, of Dedekind is like the reappearance of the monad as an abstract translation of the 'punctual turgency', or geometrical increase of numbers figured by the Pythagoreans.

It is interesting to recall some passages of Dr. Raynaud de la Ferrière taken from his *Psychological Talks*, on the "Mystery of Numbers and on Parallelism of the Tradition", where he states:

> "No meditation is possible on the number ONE; the movement begins with the TWO, when, at the level of ITSELF, appears the consciousness of ITSELF, which is in relation with the Ternary by the action of Understanding, of the Thought, and the Conscience.
>
> ONE is not considered as a number; it cannot be a figure. Symbolically it is an indeterminate center of a circle where the circumference is not in any part.
>
> The TWO cannot be a figure. It would be the diameter of a circle where the circumference does not exist, that is to say, a right line of indeterminate direction and of indefinite length, which cuts the plane in two zones. Yet it expresses the unknowable, the abstract duality, balanced and static.
>
> THREE is the original central point *whose development follows the Trinity 1, 1, 1, and* its stepup to 111 (one hundred and eleven) puts it at the center of the numeral science. It is the sign of Power, of Absolute Being, and it is called the POTENTIAL NUMBER.[8]
>
> "The four realizes active duality. It is the circle placed in rotation by the cross. The Square of two, it is the

first exaltation of the duality, the archetype of the involution in potency of being.

"One also knows that the tetrahedron (solid of four faces) engenders the cube, the octohedron, the dodecahedron, and the icosahedron. These five Platonic solids symbolized, in ancient mythology a total and integral Understanding, which entails identification with the Unique, (only) where creation has been revealed in itself.[9]

"Seven is a helicoidal number, such that tracing an ascendant line in the form of a spiral on the surface of a cone, divides the spiral into six pieces. The seventh will result from the advancement of AB, capitalized in the space during the unrolling of the other six.[10]

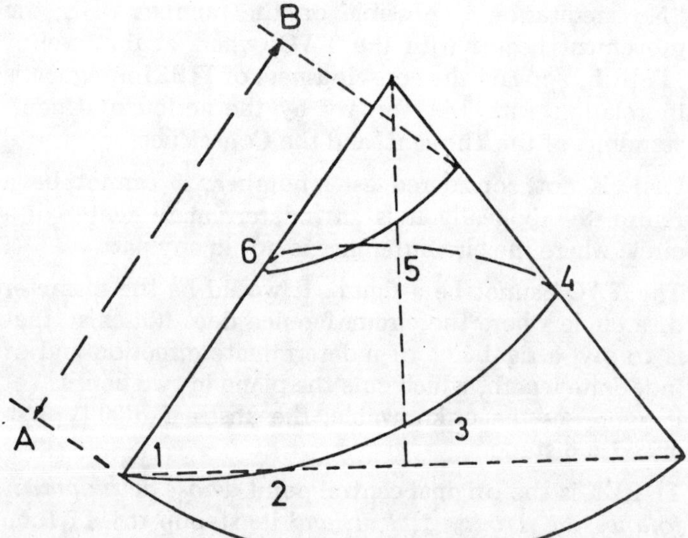

"In every *enantiomorphic* complex, the distance to the sine of the reflex-reflectant is always a multiple of 9. The function of reducing to nothing then, is that everything is completely contained in this number and its multiples.

"The sexualization of a number is the operation which consists in separating the two affinitive parts, one from the other. The problem of sexualization of numbers is the problem of Involution-Evolution (decomposition of numbers and their secret values into affinitive parts, whence originates their internal gravitation, that is to say, the permutations which allow the recombinations).

"Three hundred and sixty (360), apart from its potent symbolism by the degrees of a circumference, is the characteristic of the last letter of the Hebrew alphabet, 'TAU'. This is the sign of reciprocity, the image of all which is mutual and reciprocal.

"Let us recall that there are 24 whole divisors of 360, but as neither 1 nor 2 correspond to the geometrical figures, 22 regular polygons are then left to be marked on the circle: first the triangle, then the square, the pentagon, the hexagon, the octogon, and successively, in this manner, the polygons of 9, 10, 12, 15, 18, 20, 24, 30, 36, 45, 60, 72, 90, 120, 180, and 360 sides. To each cypher corresponds a Hebrew letter. These cyphers are esoteric values and not the number of order employed at times, which has an esoteric character corresponding to an ordinal signification.

"We might close the chart here on the symbolism of numbers saying a final word about Secret Values.

"It is written in the Sepher Yetzirah (Chapter IV, verse 4): By the 22 letters, the Creator has traced, carved, multiplied, passed, inverted, and formed the creatures, and everything which will be created. And in what manner has he multiplied them? Aleph with all things, and everything with Aleph; Beth with all things and everything with Beth; Gimel with all things and everything with Gimel. Everything turns in a circle (the Seal of Solomon). And it happens that these arise by 231 doors, and it turns out that all the words come under a similar name. It is then the number of combinations which can be gotten from the 22 letters, two by two, or said in another way: $\frac{21 \times 22}{2} = 231$. The division of the Secret Values is found by multiplying a number (for which the Secret Value is required), by the number following it and dividing by 2. The simplified formula is: $\frac{N(N+1)}{2}$ = SVN. The Secret Value of a Number must be understood, above all, by the total composed of this number; that is to say, by adding to its apparent value the sum of what is hidden. Thus 6 totals more than what it usually represents: $1 + 2 + 3 + 4 + 5 = 15$, from which $6 + 15 = 21$.

"All this is solved for Secret Values of the First Order, writing them as SV, and calling them, at times, 'triangular numbers'. The Secret Values of the Second Order, represented by SV^2, are called, at times, 'pyramid numbers'; and the value of 3 comes to be 10: 5 will be 35. Example: SV^2 of 5 is equal to $1 + 3 + 6 + 10 + 15 = 35$. There are yet Secret Values of the Third Order and higher, as SV^4, SV^5, etc. Eighty-three thousand, two hundred and fifty (83,250), is the result of adding the Secret Values of each one of the two ternaries produced by the permutations of 123. That is to say, that those two give (with 166,500) a law of enantiomorphism: $\frac{166{,}500}{2}$ = 83,250. One Hundred and sixty-six thousand, five hundred (166,500) is the total of the SV of the six numbers making up the permutations of 123.

SV of 123 =	7,626
SV of 231 =	26,796
SV of 312 =	48,828
SV of 132 =	8,778
SV of 321 =	51,681
SV of 213 =	22,791
	166,500

"Finally, let us also say a word about the sign, 'zero', which was used for the first time in the Hindu treatise, *Suria-Sidanta.* Zero comes from the Arabic 'Sifr', (empty), and from this has come 'cypher' (Sifr is the translation of the Sanskrit, 'Suria'). History has not preserved the name of the Hindu who invented the zero, and from whom the Arab geometers took it, and only history can testify if the Babylonians, the Mayans, and other peoples did not already know it before. It represents in numeral science the empty space which serves as the matrix of all numbers.

"What we have said about numbers is sufficient for our explanations of present scientific philosophy."[11]

Dr. Raynaud de la Ferrière reveals such conceptions in his *Psychological Talks,* leading us to a perception where the structure, form, and rhythm, take a plane of reality on which Number is essential to the Form. It would be well to comment also that the Pythogorean Theory of musical harmony, for example, is intact and immaculately mathematical, and is maintained in the theory of proportions.

In *A Mathesis of Psychology*, he points out again the transcendence of the encounter with Pythagorean mathematics, perhaps more than with the philosophical sense, when he says with a criterion of the new age:

"Evidently the famous Pythagorean Theorem opened splendid horizons, perhaps more in the realm of astronomy than in the philosophical sense. A detail has been called to my attention: Was his real name Pythagoras? Numerous ones have shown that this name is composed of Pithon, forecaster, and Agoras, augury, attribute, and, when the words are reversed, they produce Gurus Pitria, Master of Light, Guru, which he certainly was; an instructor-type, teaching the occult by means of reaching all things. (It is by means of the Pythagorean Tables that we obtain the keys necessary for the interpretation of the Bible.) For myself that was a revelation, and by that I began to understand better this visible and invisible world, I was called to prove this understanding from my earliest years."[12]

We are with Matila Ghyka in that Plato has been the thinker who has meditated most on proportion and harmony. His hieroglyphic on the rhythm of the world soul, mathematical and musical has been rigorously reconstructed by A.

Boeckh, whose extensive studies exposed, as the said hieroglyphic shows, the Pythagorean 'tetracts', intervening in harmony.

Thus we have one of the theorems of Plato, perhaps the most classical, often called elliptical, which he includes in the Timaeus:

> "While only one middle suffices for joining (in a proportion) two plane numbers (of the form a x b), two middles are needed for joining two solid numbers."

This theorem was commented on by Nicomachus, Erastosthenes, Eutocius, Luca Pacciolli, and Albert, and lately includes Hambridge within its dynamic symmetry.

Dr. Raynaud de la Ferrière, in this relation of number with sound, resorts to the magic squares and says:

> "In effect, all numerology is in direct relation to the musical notes and can tabulate diverse squares of diatonic intervals, in cyphered notations with their relations in the seven worlds. As an example, let us cite thirds, fourths, sevenths, and thirteenths:

SOL	LA	SI	1	2	3
LA	SI	DO	2	3	4
SI	DO	RE	3	4	5
RE	MI	FA	5	5	7
MI	FA	SOL	6	7	8
FA	SOL	LA	7	8	9

Thirds and Fourths

SOL	LA	SI	DO	RE	MI	FA
LA	SI	DO	RE	MI	FA	SOL
SI	DO	RE	MI	FA	SOL	LA
DO	RE	MI	FA	SOL	LA	SI
RE	MI	FA	SOL	LA	SI	DO
MI	FA	SOL	LA	SI	DO	RE
FA	SOL	LA	SI	DO	RE	MI

1	2	3	4	5	6	7
2	3	4	5	6	7	8
3	4	5	6	7	8	9
5	6	7	8	9	10	11
6	7	8	9	10	11	12
7	8	9	10	11	12	13

"The names of the musical notes arise from the first couplet of the Hymn of Saint John the Baptist:

UT quent laxis
REsonare fibris
MIra gestorum
FAmuli tuorum
SOlve polluti
LAbil reatum . . .

RE	MI	FA	SOL	LA	SI	DO
MI	FA	SOL	LA	SI	DO	RE
FA	SOL	LA	SI	DO	RE	MI
SOL	LA	SI	DO	RE	MI	FA
LA	SI	DO	RE	MI	FA	SOL
SI	DO	RE	MI	FA	SOL	LA
DO	RE	MI	FA	SOL	LA	SI

5	6	7	8	9	10	11
6	7	8	9	10	11	12
7	8	9	10	11	12	13
8	9	10	11	12	13	14
9	10	11	12	13	14	15
10	11	12	13	14	15	16
11	12	13	14	15	16	17

Sevenths and Thirteenths

"The name of the seventh note is taken from the fourth verse of the said couplet which begins thus: Sancte Ioannes . . . with whose initials the note SI is formed.

"The French are the only ones who have preserved these original names of the seven notes; for the Italians substituted the name of DO for the ancient UT, and in the other countries the musical nomenclature is modified completely. I have here some equivalences:

"We understand now where all vibration is from; the numbers, notes, colors, and planetary influences are closely related among themselves.

"The ancients left us a system of calculus referring to different frequencies, which permits us to express a planet in cyphers, or to say that a color determines a planet. Naturally these relations are well-known in magic. Freemasonry is not able to ignore this principle which also is manifested in rocks, animals, names, perfumes, and symbols. Esoteric investigation demands the

	SPANISH	FRENCH	ITALIAN	ENGLISH	GERMAN
12.	DO	UT	DO	C	C
11.	RE	RE	RE	D	D
10.	MI	MI	MI	E	E
9.	FA	FA	FA	F	F
8.	SOL	SOL	SOL	G	G
7.	LA	LA	LA	A	A
6.	SI	SI	SI	B	H
5.	SI bemol	SI bémol	SI bémolle	B flat	B[1]
4.	Bemol	Bémol	Bemolle	Flat	C[2]
3.	Sostenido	Diese	Sostenuto	Sharp	D[3]
2.	Mayor	Majeur	Maggiore	Major	Dur
1.	Menor	Mineur	Minore	Minor	Moll

1. (in German, in this one case, Si bemol is in need of support)
2. (in German, this syllable is added to the letter of each note, for example: C for Do bemol)
3. (this syllable is added to signify Sostenudo: D for Resostenudo.)

most varied skills, and above all, the immediate comprehension of the relations and laws of correspondence between the different elements of the Kabalah, astrology, magic, architecture, and music, so that in synthesis, the correlation which exists between the diverse realms of science, art, and philosophy may be established.

"At once, it presents us a study which derives the formulable expression of the human heart itself:

Further on he makes mention of Archeometry as 'the instrument of evoking the past, necessary for construction in the present, as a means of synthesis and of regeneration of all the intellectuality in the future.' And he adds:

"A system exists which not only can join all the known concordances, but can also serve to decipher mythologies, legends, religious principles, etc., through an architectonic method. Thus we find ourselves in the presence of genuine possibilities for continuing the Oriental, Western, Semitic, or Aryan traditions.

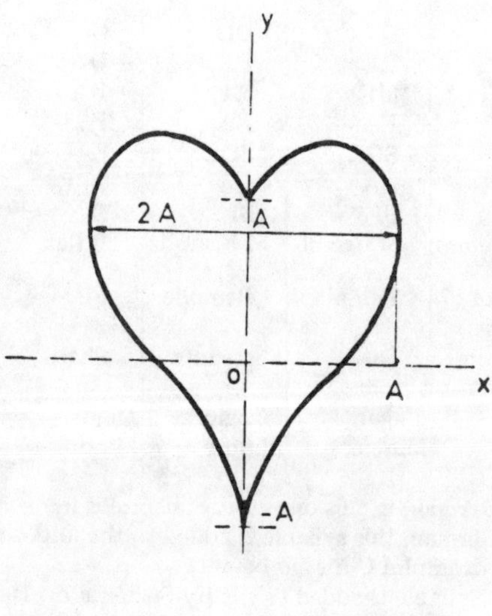

$$Y = \sqrt{x^2} \pm \sqrt{A^2 \quad x^2} .$$

"This system is Archeometry, which is truly the most elevated science, because it permits broaching with facts, whatever kinds of problems arise as far as true learning, thus offering the possibility of rectifications so frequently necessary in the history of human education.

"As an instrument of outlining it has, nevertheless, the advantage that if each guild of artisans has its proper tools well-defined, it serves also for all other uses; it is the key of sonometry that adapts the colors to the paintings, and at the same time the forms to the architect. One does not try to verify with it the pre-established instructions, but rather, to investigate the manner of raising expressions in accord with a defined method. It does not build anything, but it presents the principles to obey for undertaking whatever is outlined.

"Above all it re-establishes a synthesis of the arts, re-adapting expression to the true founts. It gives the key of successive religious and scientific adaptations of antiquity. It permits the architect to construct forms according to a name, an idea, a determinate color, respecting the vibrations emanated from a harmonic curve, because they correspond to a planetary influence in relation with the origin of revelation.[14]

"So, then, it becomes easy to notice the alterations which a symbol has undergone, since Archeometry is represented as a special planisphere. This expresses to us a graph of the science of the cosmobiological and morphological correspondences, based on the word, and on its equivalents.

"Naturally this science can be applied to all esthetic duties. Nevertheless, the monumental synthesis has been represented by the religious edifice, and it is that to which we are going to devote a paragraph.

"The harmonic laws of proportions, of the forms and music, are the same. Thus, a sacred object, a column of a church, a lodge of initiation, a temple in its totality require a metric synthesis which evokes the name toward which the aspiration, the symbol, the offering of the ideal, etc. . . . ascends.

"Mathematics, geometry, physics, and astronomy perform a great role for those who want to reform the world; they are solid concrete bases for establishing first in the material world, before the elaborating of a better spiritual universe.

"Archeometry is a material translation of the Word, in form, color, perfume, resonance, and taste. It is a figure of synthesis in a circular form. It is divided into concentric zones which contain the correspondences of names, letters, notes, zodiac, planets, alphabets, (Wattan, Hebrew, Latin, Samaritan, Chaldean, Assyrian), and vibrations as a metric model which reforms Sonometry and serves as a rule for tracing the esthetic proportions of all graphs.

"The primary colors, yellow, red, and blue, are distributed triangularly within the circle, in such a way that theoretically the color white is found always to exist at 180°, opposite each one of these three colors. In this way there results 360 definite tones. Each one of them is identified with a number of an order which permits rapidly discovering its exact composition. Certainly, the relations of music are inscribed there, according to the physical method, inversely proportional to the model of the current middle: (D-flat = 100,000, or 1 meter; the cypher 100,000 = 625 x 160). The tracing of a form, whatever it might be, thus comes to be an expression of the key of a universal canon. That is to say, that according to a lone name, or a single idea (equally a color or a sound), it becomes possible to obtain a pattern, a graph, an entire building, with the exact proportions of breadth, height, and so forth. Thus, it already deals with an esthetic monument, a religious construction.

"With an inverse procedure, equally adapted to the literature of colors, forms, etc. . . . this intimate concatenation of words, of ideas, of ancient sciences, and of symbolic studies, permits, for want of files and documentation, reconstructing any religion, philosophy, or legend, by simply reproducing on a paper or a sheet, the constitution of the heavens. Having once the numerical value of the words, the confusion of the languages matters little. The letters are preserved in the manner of their original forces, that is to say, with the sacred idea. (MA in Archeometry is equal to 40, and in veda, wattan and numerous oriental languages they say WATER; however, in the extreme West the Incas pronounce ATL for signifying this same element.) That which is important is the intrinsic value and it only counts in the final result."[15]

Likewise in Egyptian art we see the proportionality of volumes and forms. The hieroglyphics still seem like a universe of harmonic perfection, premeditated and realized, says Ghyka. He considers that from the point of view of its pure form, the work of art can act on that which contemplates it in two different ways for producing pleasure or even the ecstacy. As Ruskin said, art is adoration: through a minor harmonic force, it is hedonistic; through organization of the chaos of sensations (with sympathy, Einfuhlung), it is understanding, according to the contemplative spirit, with the proportion or symphony which the artist has produced. If the adoration is through enchantment, in this case, rhythm can be sovereign, or it is by means of the emotive wave of what contemplates or listens, or it is with the interior rhythm of the artist, which in its time is the reflection of the pulsation of the Great Rhythm, the Great Harmony, and in this action of enchantment as in the rest of artistic expressions, there is a law of number which organizes.

Still more, this law regulates and balances it, observes Le Corbusier, in the visual consciousness which automatically makes the necessary optic corrections, such as rectifying the retinal images.

Pius Servien defines rhythm as perceived periodicity, and d'Eichtal considers that rhythm is in time, what symmetry is in space.

For both the Pythagoreans and Plato, mathematical philosophy was the entire philosophy, and they included within it music with its numerous elements and discontinuous rhythms. The thought of Dr. Serge Raynaud de la Ferrière on the mathematical universe involves the extraction of fundamental concepts for a new age, qualified by that Cosmic Biocybernetics which is Cosmobiology as eminently Uranial and Mercurial.

Next to the images of a universal wisdom which stretches a bridge between two points of the world, between two precessional eras, there are transcendental images in the background. By even the slimmest ladder, they lead from one era to the other, a ladder which is rocked unceasingly in the Thought of the Master and invites us to lift our spirits: it is the mathematical ladder of his Universal Thought!

NOTES

1. Mathematics in the Behavioral Sciences. pp. 55, 57.
2. The Great Messages. p. 120.
3. Computers and Computation. p. 148.
4. ibid. pp. 14, 16.
5. Perroux Interrogates Marcuse. pp. 121, 123-4.
6. The Golden Number, I. The Rhythms. p. 22.
7. ibid. pp. 23-4.
8. Psychological Talks. I. p. 86.
9. ibid. III. p. 125.
10. A Mathesis of Psychology. p. 474, fig. No. 65.
11. Psychological Talks. Tome III. pp. 128, 140, 142-4.
12. A Mathesis of Psychology. p. 33.
13. Black Book of Freemasonry. pp. 93-5.
14. ibid. p. 99-101.
15. Black Book of Freemasonry. p. 103.

CHAPTER 7

The Mathematical Cosmovision of the Future

We have presented so far some parts from the panorama of the Master's thought that show the field related with mathematics. We have seen the considerations which allow us to establish a historical framework with a conceptual atmosphere. It is this in which his living experience has arisen which occupies the theme of this work.

Nevertheless, within the reduced space of this work, there are many aspects, relative to this theme in the work of the Master Raynaud de la Ferrière, like his geometrical applications, analogies, the relation with the alphabets, the manipulation of the numerical values, and the mathematical models which need to be incorporated parallel with his conclusions about the equinoctial precession and celestial mechanics. These last have been related in the *Scientific Theory of Cosmobiology*; still they also require investigative application at the theological and sociological level.

We have had to consider the past as well as the present in the intellectual status of mathematical thought. The present evolution makes it necessary to include a mathematical prospective of the future, especially that of Dr. Raynaud de la Ferrière since he outlined the reeducation of man in a world united by learning. He focused on the future of a new age, begun in 1948.

The thought of the Master is eminently the thought of a new human era; the previous era is finished. Unity with the Great Ancient Principles is the basis of a dearly described new age: the new Aquarian Age as a biocosmic imperative of computable precision.

The mathematical cosmovision of the future brings us to the precessional eras, and to the astronomical proof of the topocentric limit in our biosphere with all its individual and community consequences in human society. It also brings us to a proper understanding of ourselves in the more intimate substrata of our organic and thinking life which computer science reveals.

We could begin, for example, with Mary A. B. Brazer, an English neurophysiologist at the University of California. She describes in her article, "The Analysis of Brain Waves" (from *Mathematics in the Behavioral Sciences*) that to analyze the electric activity of the brain, the computer extracts significant signals starting from the electric 'noise' originated at the back of the brain; this noise under normal conditions, makes indecipherable whatever the register obtained. The computer becomes more useful still when it functions in 'real time'; that is to say, while the subject being observed is also connected to the recording equipment. This permits the investigator to modify the experimental variables during neurophysiological studies.

In man, says Mary A. B. Brazer, the difference of power flowing between distinct points of the hairy unshaven skin oscillates between 50 and 100 millionths of a volt, which is the tenth part of the electrocardiogram potential. These waves are especially prominent in the back part of the head, in the cerebral zone corresponding to the areas of visual association; the waves registered here receive the name of alpha waves. The alpha rhythm, which has a frequency of 8 to 13 per second in the adult subject, is more conspicuous when the eyes are closed. Upon opening the eyes, the alpha wave disappears momentarily.

Owing to the regularity of the alpha rhythm, its frequency characteristics were much studied in the first days of the electroencephalogram. Physiologists thought that if these waves were analyzed mathematically, they could perhaps discover certain components which remained hidden from simple ocular inspection.

The experimental material of biology, continues M. A. Brazer, and especially that of electrophysiology which studies the brain, is a living and changing material, and the investigator has to avail himself of the opportunity of extracting from it all the information possible before the passage of time induces new variables. The computers which are used currently in industry, and in businesses, have not been projected absolutely with those ends. It is an imperative of the future.

She continues, saying that H. W. Magoun (now at the University of California, Los Angeles) and G. Moruzzi (now at the University of Pisa) together discovered in 1949 that the reticular system has a fundamental relation with the state of alertness of the organism, and with the intermediate shades of behavior between the state of permanent vigilance and the forgetfulness of dreams. Later works have assumed as manifest other models which can only be identified, in the electrical registers, through the fine analysis which computers are capable of realizing.

Belmont Farley, of the Lincoln Laboratory of Massachusetts Institute of Technology has developed programs for the investigation of the alpha rhythm trills which appear and disappear in the human EEG, and give indications about the level of consciousness and the normality of the brain.

One of the first investigators who encouraged the use of electroencephalographs for exploring this approach was the mathematician Norbert Wiener of MIT. His great influence was noticeable in much of the work done through computers.

No exposition of the use of computers in electroencephalograph work could omit its application to the studies recently done concerning the correlation between the superficial activity and the depth of the different parts of the brain. What is the correlation between the graphed waves starting from the exterior of the brain, and the interior activity? asks M. Brazer. With what surety can we affirm that an EEG is 'normal' if it is obtained only through the electrodes situated on the hairy skin?

While more and more clinical investigators are adopting computer techniques concludes our author, it will be possible to adapt to the electroencephalograph—which can only obtain data starting from the hairy skin—a compendium of correlations for assigning probabilities to the events which take place in the hidden depths of the brain.

In the same collection of articles, Patrick Suppes, Professor of Philosophy and Statistics at the University of Stanford, thinks that the truly revolutionary function of computers in education resides in the new field where these machines assist in instruction.

Let us predict, says Suppes, that within a few years, millions of children will have access to what Alexander, son of Philip of Macedonia, considered a royal prerogative: the services of a tutor as well informed and sensible as Aristotle.

The most powerful argument, he adds, in favor of instruction through computers comes from an ancient concept that refers to the advantages, partly proved, and partly conjectured, of individualized instruction. At the university level, individual instruction under tutors has been one of the glories of Oxford and Cambridge. The criticism directed in modern times at this method does not refer to its intrinsic value, but more to its economic inefficiency. A sufficiently broad agreement exists that the greater the degree of adaptation is of the scholarly program to the individual students—each one of whom possesses a capacity, rapidity, and 'style' of suitable apprenticeship—the greater is the possibility of success in teaching.

To go on describing the innumerable possibilities for the application of computation would bring us not only to mentioning the journeys of satellites and interplanetary laboratories, but to considering the future of computation in the social function of Cosmobiology which we have outlined in the first compendium of the *Scientific Theory of Cosmobiology*. There the biocosmic controls of phenomena to the second, by means of the mathematical models of Wendel Polichian topocentricism applied to the cosmobiological social facts, allow the foreseeing of social and collective conduct, as well as an optimum curve of possibilities in the application of its mathematical basis for its social function.

As we say in the *Scientific Theory of Cosmobiology*, (an edition supported by the Directorate of the National University of Trujillo):

"The practical cosmobiological applications of anthropology, between the investigator and the people, for the well-being and health of remote communities, isolated, and in rural zones, where all in the world live, facilitated social adjustment from within and without, as the decisive unifying element, integrated par excellence. This will be able to help in a great part, the socio-cultural diagnostician of a forgotten community, margined barrio, young city, etc., though some accustomed forms are required for understanding socially a community, (education, social mobility, economic level), costly and prolonged investigations. It saves time and expense, at the same time that it furnishes data that are either slow to obtain—like the intellectual capacity and vocation of the population, the activities and works in which each one excels, for helping them thus to arise—or they are facts which, by being subjective, are not really indicated in the population.

"The cosmobiological facts diminish the problem which arises when the community adopts a state of alert or a negative attitude toward the investigator. It can avoid this, in great part, and make indirect questions for knowing attitudes or personal matters. Knowing cosmobiologically the ground he treads, it will be easier for the investigator to help the community on the social plane.

"Their unities of time will be useful to the interaction of small and related groups, interactive with the indispensible prediction of behavior which will permit the dynamic configuration of dynamic models of culture.

"Social scientists can study an interchange of tendencies and pressures elaborated and coordinated in a cosmobioclinic of conservation of human resources for evolution in the sociocultural systems, even when it implies a complete readjustment of the basic theoretical suppositions.

"In Cosmobiology and in its methodology of investigation as much of the data of the mechanical system as is persistent will be called forth. This takes into account that sets of interaction will be based on data of precision as many times repeated as the interaction of sensitive axes and the frequency of planetary positions permit it. They will be able to be defined in mathematical language for their multiple interpretation and catalogued through diversified levels.

"The cosmobiological units will be applied through the socio-economic, as André Barbault demonstrated in his chronological statistics of the cyclic factors (units of time of socio-economic phenomena). The investigation of the environmental factors of the socio-economy is extended to the biocosmic field. The environment will not only be socio-geographic, but also sociocosmic. The socio-economic cosmograms for use by statistics, will permit important and different verifications and modifications in the face of new conditioning factors, and their investigation will turn out beneficial for describing the social function.

"In the judicial sciences, the *units of time are verified*, for the elucidation of interrelations between criminal and victim, for the study of extenuating and aggravating circumstances of culpability in the judicial process of criminal cases, for the diagnosing and evaluation of the regenerative stages of those sentenced for long terms, and for the prognosis with statistical base of their reintegration into the society.

"The Cosmobiological units of time are indispensible in developing a vocational diagnosis for investigating the educational capacities of students, likewise for presenting charts of groups of homogeneous students, not only in their type, but in units of time of greater aptitude and yield. These permit the choosing and foretelling of the university course or work duty through stages of development and objectives.

"It is most convenient for cosmobiological interest to make the first mention of the congenital malformations and of the problems of childhood at the moment of birth.

"Medical investigation on congenital malformations in the instant and place of birth, furnishes data for the planetary kaleidoscope, as Dr. Raynaud de la Ferrière calls it (The Great Messages, 159, where it is treated in the respective chapter). In the field of congenital psychological alterations and the criminogenic factors of infantile delinquence and mesalogical factors of the community in the youngest ones, this opens the recourses of cosmobiological application. It does this equally in the rehabilitation and protection of the mentally retarded, in the orientation of the mentally handicapped, in the services of special education, and in the study of scholarly failure. Even for the drug problem, the parameters of cosmobiology show their validity.

"a) To the criminogenic factors which cause crimes, Cosmobiology can join, with a good percentage of priority, the congenital psychological alterations which appear clearly in the instant of birth with geographic latitude and longitude (place is the exactitude). As this study is kinetic, it permits establishing predictions which serve as guides for the conduction and pre-symptomatic formation of the potential delinquent. This regeneration option, affirmed on cycles of congenital predisposition, is equally valid for the pre- or post-criminal stages, with the object of studying their potential for rehabilitation of the acute tendency. The diagnosis and the prognosis include valuable observations on the criminogenic factors to which the psychiatrist Hugo Meneses has made reference, (anxiety, schizophrenia, manic-depressive psychosis, neurotic personality, inferiority complex, mental conflict, and other abnormalities).

"b) The mentally retarded person can be diagnosed and prognosed in order to define his potential compensative abilities.

"c) In special education, the adaptation of the child can be cosmobiologically planned by stages modified according to well-established lapses. The greater yield of his abilities and predispositions can be seen, even in the most private of possibilities, as well as for the stages, previously determined in time, of the variation of the favorable environments.

"There must also be established the study of the trinomial, parents—mentally retarded child, and the polynomial, father—brothers—mentally retarded child, for orienting his warning stages, and for optimizing family interrelations.

"d) As for the scholarly failure, studying biocosmic factors allows the drawing up of a plan for changing motivation, or educative goals, including convenient periods of recess in the study. Convenient times for intensifying or regaining the lost study, can be considered as well as prevocational or vocational orientations so that the study harmonizes with the person's potentials, inclinations, and aptitudes.

"e) In relation to the disadvantages of a child, the cosmobiological analytical chart outlines the disadvantages, analyzing their time variables and compensating aptitudes in the sectors or houses of the sensitive axes, or cusps.

I — the inclinations or congenital malformations. the form of directing the others.
II — the economic postponements.
III — the inability for studying and expressing oneself.
IV — the family frustrations facing the father and mother or the environment of his house.
V — the inhibitions of creativity and the requirements of recreation.
VI — the problems of health.
VII — the repressions and differences of the environment.
VIII — the lack of gifts and rewards.
IX — the frustration of traveling or segregation by religion or ideology.
X — the limitation of social position.
XI — the displacements of help and protection.
XII — the disadvantages by isolation.

"f) With respect to the tendencies and difficulties which sociological study reveals for some educational problems, the more transcendent conclusions of Dr. Raynaud de la Ferrière belong to the receptivity of the sensitive axis of the vernal point. They are biocosmically determinant of the present stage of transition between two eras, in which an inversion of increased multiples is performed toward an attitude of searching for the wisdom of social equilibrium, and of the return of man toward himself, putting the individual at the service of the community. The study of this intersection brings us explicative parameters of the standards of achievement for the active transformation of the models of the world, embracing the mentally retarded human beings and those who are found in social or congenitive disadvantages.

"The answer of Cosmobiology is pragmatic, transformed, and evolutionary for these great problems of childhood and other systems of education.

"In the service of cosmobiological orientation in the field of the rural, laboral, worker, or professional application, they are applied with a precision of time in the cycles of work. The norms of possibilities and variables which the worker has within his own experience as well as proper foresights of psychology and preventive medicine and reeducative standards for knowing how to live together, besides other perspectives are applied through this.

"Thus we can now observe some ways Cosmobiology is employed in social assistance.

"First—orientation of the laboral vocation in variables of time and vital stages for the personal realization and behavior of the incentives and faculties in the specialized work, in the non-specialized work, in the farmer, in the employee, and in the professional. Precise control shows the possibility of forecasting and understanding through the special study in the worker's case, the diverse branches to which he can aspire in a similar industry, and matters which broaden his general culture, according to the cosmological modules of time. The possibility of studying in adequate lapses is considered, as well as the foresight of stages of a laboral environmental change convenient for the preservation of his development as an integral man.

"Second—as for laboral psychology, the units of cosmobiological time are equally applicable aids to the psychotherapist and to the foresight of crisis, of the experiences of humiliation, shame and fault which do such damage to the development of a healthy ego. They are also applicable to excessive emotional participation, by knowing when to control, to discipline, and to impose clear limits on its impulsive communicative answer in a form which does not turn out destructive or inhibiting. They are the cosmobiological stages which bring us to the study of privation or sensorial enrichment and what we know about the powers of self-comprehension and self-control, with its basis in a rhythm and a reeducation for knowing how to live together.

"As the psychiatric Doctor Gardner Murphy says in the interview already mentioned that he made with his colleague Willard B. Frick, 'I believe that psychology of the next thirty years will be very rich as far as scientific self-discovery type'. We can interpret the previous messages of the viscera and the muscles in a form each time more precise and we can thus understand the basis for a rhythm and an order in life; and understand the marginal messages of the vital organs which are related with our esthetic experience, like the rhythms which I mentioned, in the answers to cosmic order. I believe that it could be shown that the ecology in the ends of physics, chemistry, geology, climatology, and meteorology, to which man is adapted, introduce in his body different rhythms, and different supports, in different parts of the world, and in different moments of history. Thus, I do not believe that there is some mystery in the fact that man is isomorphic with respect to the cosmic structures. The phenomena can be observed up to the point in which it is proved that we are not facing the whims of different media, but the basic uniformity of the mathematical support."

"Likewise, in the interview which the same Willard B. Frick had with the psychiatrist Abraham Maslow, he said; 'There is something more than self-realization, more than the full identity of the real person, and I believe that I would go much farther than Frankl and would speak of the 'cosmogenical' as someone has called it."

"Third—for the childhood of the worker and the farmer, it helps in the infant formation and the prevocational determination in the quality of precocious diagnosis of inclinations and the development stages of faculties, from the age of the binomial, mother-child, the important study of the inter-individual cosmobiological relations. This model of investigation which serves as a basis for the cosmobiological application in all the relations between individuals, points out the particular reactions of each mother-child case, and the time units in which they are modifying the motivation and reaction of these relations. This permits a more profound orientation, as much for the formation of the child from the conscious action of the mother, according to the process of modifications in both, as for its better comprehension and solidarity. Thus it is possible to orient and remedy many family and social difficulties that are the result of error and the bad management of interpersonal transactions, of the predominance of supports of permanent crisis, and the inauthenticity of behaviors which at times come to be manifested in the absence of creativity, lacking originality or the force of expression.

"Fourth—as for those young people, or those in the slums and areas of misery, the cosmograms of the small and the great communities, by means of the individual data, permit an orientation in the social diagnosis and the mechanisms of integration, and thus a better yield and orientation.

"Fifth—those levels of application are transferred to the field of the qualified employees, office clerks, bureaucracy and those in the professional sector, in their respective environmental requirements. The capabilities are enormous, proper to a cosmic biocybernetics at the level of all people, for the betterment of their use of free time, and for making them capable and improving their yield.

"Sixth—to promote cosmobiologically the cooperation and the family and community participation in the educative and reeducative work.

"Seventh—to provide adequate services of orientation which are projected in the benefits of all the population.

"As an instance of planning for the participation of Cosmobiology there is a wide field of medicine, preventive and social, which as a discipline does not have precise limits, since those vary between the Europeans and the Latin American countries, and are subject to continuous change.

"After setting out some projections in preventive medicine, we could not pass over the frontiers of social function without considering the didactic. This is done not just in its socio-pedagogical framework, but in the measures and organizations which have made possible the pedagogical references, to which by their transcendence, the chapters of Part VIII are dedicated.

"The first problem of Cosmobiology has been, for a number of years, the impediment of the numerous mathematical controls for its statistical basing that has been produced through investigation and computation. The next problem to resolve has been the educational application for which it is extended without the purpose of gain, through the specialties or by orientation of an interdisciplinary base, to the zones of great or small nuclei of population.

"Thanks to didactic it is possible to bring its educational application from the heights of scientific investigation to the level of education for the illiterate masses of the population.

"Sociocosmobiological educators will be able, and ought to be, professional educators, union directors, literacy developers, volunteers, broadcasting MC's, journalists, and sports trainers.

"The pre-vocational and vocational cosmobiological diagnoses allow fitting the worker, the student, and the professional with the education which they ought to apprehend and acquire for the optimum determination of their abilities."[1]

Cosmobiology implies the computation of simultaneous controls on forty static points called radicals, the same static points from each annual cycle of the life of man in the case of control of the phenomena in the human species. Likewise other combinations are made with the same 40 factors in dynamics of different velocities, and repeated by other movements of declination and obliquity in constant contact with the 16 angles of sensitivity. All this represents an intensive multidimensional work in the processing of data in real time, that is, while the events are taking place.

Thus, Cosmobiology comes to constitute an advanced branch of the mathematical cosmovision of the future.

Looking to the future, cosmobiological computation makes us ask: At what speed can one expect to operate the computers in the cosmobiological control of everything in the analysis of one community? What will be the cosmobiological 'hardware' of the future? What perfections are possible in the forms in which the components of a computer are organized for effecting the work of controlling an interrelated community with social, pedagogical, and psychological objectives? Or in the comparison through static schemes and cosmobiological kinetics, of the simultaneous behavior of distant communities?

Still, when cosmobiological computation can be satisfied in the studies of investigation by an individual with a computer of the type, K24, with respect to its social function in communities, surely in its future perspective, it would require a line of advanced machines. Perhaps they would be like those conceived and developed at the University of Illinois, the ILLIAC IV, for example, which is actually a set of 64 ordinator 'slaves' capable of executing 100 to 200 million instructions per second. These are capable of performing 64 simultaneous calculations, transmitting directions to individual memory banks from processing elements, and controlling a program while it registers a certain number of processing elements which simultaneously executes the same instruction on different data in each memory of the processor. The time of access for extracting a number from the memory is 188 nano-seconds, and owing to the additional logical circuits, this is increased to 350 nano-seconds.

The additional data can be deposited on a rotating magnetic disc whose writing mechanism is a laser ray, with a capacity of a billion bits in 128 tracks each one possessing its own head for reading and writing, with a filing memory of a trillion bits.

It is curious and significant that this formidable force of design and of programming and correction has been described in "The Most Rapid Computer" from the *Scientific American Selections*, by Professor of Computer Sciences and Director of the Center of Advanced Computation of the University of Illinois, D.L. Slotnick (Doctor in Applied Mathematics to The Celestial Mechanics). Therefore, a good part of the basis of Cosmobiology, with its mathematical models derived principally from the astronomy of position depends on this latter specialty. This great design is achieved under the direction of Arthur B. Carrol.

John Von Neumann, one of the first computer theorists, considers the numerical prediction of time as one of the most important motivations for his work. Cosmobiology as a science of time fashions mathematical models which predictively trace the phenomena to the second through the future years of the process of a clinical history for early diagnoses. These adapt the 'when' of the incidence of a pathogenic relation which is observed in a symptomatic framework in a previously given fact, thanks to the well-aimed calculation of predictions.

We would say that in a certain manner the Cosmobiological computation will participate in the future in what Steven Anson Coons (Professor of Systems and Sciences of Information at the University of Syracuse, in his article, "Computers in Technology", also published in *Selections from Scientific American*) rightly calls, introducing profound changes in that broad section of technology which could be named 'creative engineering'. He mentions there J.C.R. Licklider, of IBM, saying that there is a symbiotic relation when the terminal and the man are bound by a relation in which each part can bring to an end its individual potentialities in a common undertaking when the deficiencies of one compensates for the other. And he adds:

> "Man is in his element inventing and organizing ideas, making associations between notions with apparent relation, recognizing configurations and suppressing unimportant details; man is creative, unpredictable, at times capricious, sensitive to human values. The computer is almost exactly what the man is not. It is capable of giving attention to an unlimited quantity of details; it is immune to distraction, precise, and sure. It can execute easily the largest and most intricate calculations without a mistake, and in less than a millionth of the time which its human collaborator would need. It does not experi-

ence emotions, or at least we do not suppose so; it does not suffer from tedium, nor from fatigue. It is necessary to instruct it only once. Starting from that moment, it remembers perfectly until you tell it to forget it, in which case it forgets instantly and absolutely."

As Anthony G. Oettinger, an investigator in the program on technology and society at Harvard University, in his article, "Computers in Science" (likewise published with the previous ones in the *Selection of Computers and Computation of Scientific American*) points out: when the computer performs in the field of scientific applications not only the role of instrument, but that of actor, as active participant in the development of scientific theories, this is the field in which it promises to make its most profound impact on science. Just as Isaac Newton reduced the argument of celestial mechanics to the compact writing of the differential equations, Urban Leverrier and John Couch Adams rested the dialogues with their lengthy and detailed calculus on an enormous mass of astronomical observations. Johann Galle and James Challis aimed their telescopes toward where the calculations said that they ought to do so, and they discovered the planet Neptune.

The Hungarian Professor of Mathematics, Wendel Polich, has demonstrated through his topocentric discovery, in which he was aided by Nelson Page, that in Cosmobiology the determination of a phenomenon to the second in human life, in the organic, or in the everyday events, permits 'discovering' the precision of the ascensional longitude or declinations of a planetary point, or vice-versa.

In such cases the language of the programs is the language of the sciences; the program is the original and only version of the argumental standard and not simply a translation of mathematics.

Thus, in the mathematical prospective, we could go on considering into the future, a universe of immediate or mediate applications, which not only start from the advances of man but that return to him with the impact of his modelling. The environment turns into nature; mathematics turns into the living and creative environment, which amplifies the notion and the application of the mathesis, mathema, which Plato called 'The greatest good'.

NOTES

1. The Scientific Theory of Cosmobiology, Acts of Sciences, I. pp. 47, 49, 58, 60, 62-3, 65, 69, 72, 85, 94.

PART III

THE MATHESIS

CHAPTER 8

The Implication of Mathema Is The Mathesis—the Qualitative Change of the Dialectic Method

Dr. Raynaud de la Ferrière, through a dialectic outline of the mathesis, extends the application of mathema to what Plato called "the greatest good" (*Republic* VI, 505a), that is to say, all that is the object of the acquisition of knowledge. Sextus Empiricus thought that the mathema implied, in addition to the thing learned, the one who learns it and the mode of apprehending it, and applied the name of mathematician to all those cultured in science in addition to the philosophers. The two terms nearest to mathema are: 'discipline' on the one hand, understood as science in what is learned and taught; and, on the other hand, 'mathesis'.

Mathema is a science-like discipline with regard to all that which is the object of the acquisition of knowledge, and which Plato called in the idea of the good, 'the greatest mathema'. Kant exalted mathema to the propositions of the mathematicians which are obtained as a construction of concepts. Dr. Serge Raynaud de la Ferrière's conceptions of God through mathematics, concerning the divinity as the Supreme Cause which contains the non-manifested effect and the manifestation, as life-form-thought, achieve the quality of mathema. Mathema also entails study, research, and analysis as a discipline, with method and logical order; it is not the uncontrolled and immoderate attention to study. It is not disordered study without method and logic which makes ideas jump and wander. It is the systemic scientific discipline of structurable organization. It is practical knowledge for higher learning in the fulfillment of man.

The other implication of mathema is mathesis, whose special importance in this work is the fact that Dr. Serge Raynaud de la Ferrière categorizes it in three great aspects: the culmination of a qualitative change in the dialectic which he proposes: thesis-antithesis-synthesis-mathesis; research into the living synthesis for the integral fulfillment of man as a universal identification, and finally, one of his principal works has been entitled, "A Mathesis of Psychology". Its first part is the comparative experience of himself as western man, as mathematician, as physician, as scientist, who through the modern scientific resources transcends the limitation of Western canons and progressively penetrates the Identification with the Manifestation, with the non-manifested effect and the Supreme Cause. In its second part it is his definite eastern expression typically Hindu, which fulfills and rectifies the systems of the yoghistic wisdom and achieves Yug in the Identification with the Supreme Cause. He himself defined this in his vision of an epistemology of life, which he considers necessary for focussing the analysis of the diverse manifestations of that new concept of divinity, fitting the process to his own psychological evolution.

In these aspects of the term 'mathesis' we cannot overstep the framework of this book, and we will discuss it only in correlations with the theme of his thought which we are treating. In other works of this series, entitled *Serge Raynaud de la Ferrière, His Primordial Thought*, we will refrain from what develops the explanation and analysis of his dialectic and of his scientific conclusions, or of other aspects of the evolution of his thought.

Let us look now at the mathesis. This term whose Greek origin (according to the encyclopedias: German etymology of J. B. Hoffman, and the English of Webster, Alexander, and Chalmers), originally implied mental discipline, research, acquisition of knowledge, and effort through exploring. Etymologically it is allied to the Greek word Mathein, to learn, and Math learning.

Leibnitz calls Universal Mathesis the combinatory art, or the universal characteristic.

It would be fitting to think here on the knowledge of 'simple natures' in Descartes' "Rules". This creative intuition, or rather, 'constructive', in the sense that it deals with a spontaneous activity of the subject without which no essence could give itself up to thought, requires, for example, effectively realizing an act of reflexive attention for capturing the essence of the act of 'perceiving' and recognizing the presence of such an essential invariable in the singular acts of perception.

So then, the true reality in matters of knowledge is the epistemological act. This always acts upon the product of something previous to correct it, so that at the end of its work of eliminating all traces of image, of visual residue, the true science of knowing has produced a new form. To insist on this creative and almost demi-urgical character of scientific activity is to make of it all one work.

Another significant etymology of the root of mathesis: Matha, according to Yves d'Alveydre is 'the reason of reasons, the incidence of reflections, the legalization of laws, the eudoxia of doctrines.'

Protosynthesis, based on Cosmobiology allows the consideration of other constitutive criteria of mathesis.

According to Yves d'Alveydre: "Philosophical criterion: Man thinks as he lives, proportionally to the education given him by life and the instruction given him by science, which leads to the True Experience, the Truth constitutive of the Universal Experience, the reflection of life (Bios) in the Cosmos.

"They are objective criteria through observation and experience. They come also from revelation and this concordance is Wisdom itself.

"2nd Criterion: Comparative; Cosmobiology and Society. The association of the intelligent and free harmonic functions is a social expression.[1]

"Science is an instrument of supreme intelligence, of foresight and inexpressible prudence throughout what it encompasses and supposes, whose absence determines chaos.

"3rd Criterion: Mysticism as the intimate union of various criteria concerning the Supreme Cause."

In this regard, let us make an insert in the epistemological theme, concerning the epistemological application in mysticism.

Bochenski recognizes that it is necessary to face a situation more complex than that which appears in formal classical logic, for symbols must be used which are expressions of expressions, that is to say, symbols of symbols. According to Bochenski, through the discovery of these, if we cannot give exact formulations of many formal properties implied in relations used by metaphysics and theology, this is not due to the lack of such formal properties, but to the underdeveloped state of biology and other sciences, from which metaphysics and theology must extract their analogical expressions (and the contents of them).

Thus, it is not strange that within the thought of Master Raynaud de la Ferrière there exist innumerable paragraphs of epistemological considerations related to mystical observation. For example, in the series of Psychological Talks, No. XXXIII, "Mystics and Humanists" he says:

> "It is fitting to give the original meaning to the word 'mystic', which in rather abusive constant use, has partly corrupted it. The 'mystic' of science, the 'mystic' of work, the 'mystic' of the fatherland, according to Peguy, this word is in opposition to 'politic'. It is wrapped in a halo of mystery and ideal, the dry precision of the concept. It will be necessary to undo the idea of the mystery of mysticism from that of religion.
>
> "Mysticism, in the true sense, has something in common with 'mysticism' according to the current use of the term, which stresses the irrational element rather than the social, moral, or dogmatic aspects of religion. Perhaps it could be defined thus: the mysterious desire proven as sacred, prior to all rational justification, often unconscious, but profound and incoercible, of the soul which strives to enter into contact with something which it will consider as Absolute, generally its God, but sometimes a vaguer entity, such as being itself, the great all, nature, the soul of the world.
>
> "This first element, simple and identical everywhere in its essence, is evidently expressed in diverse forms, according to the medium in which it presents itself and the spiritual quality of the individual. It revives the incantations, and the dances of primitive men. It is manifested more nobly to the authority and measure which civilizations set for it: Buddhism in India, Taoism

> in China, the Greek mysteries, Arab and Persian mysticism. The great religions of the past have their mystical books. When philosophy has developed among a people, it also tries to translate discursive formulas, which, essentially, go beyond reason.
>
> "For the Christian—before the mysticism of other religions, however infinitely thrilling and respectable—babbles only, "Christ is the Only Word . . ."

At once Master Raynaud de la Ferrière sets forth other observations on mystical thought; "For some centuries past, it has been the custom to call those people 'mystics' who have had the experience of God in them to some super-eminent degree. Speculative mystical theology describes this same experience and deals with it theoretically.

About 1260 Master Eckhart of Hochheim was born. Juana Ancelet-Hustache, (in *Master Eckhart*, Paris 1956) describes the work of this great mystic.

> "The thought of Eckhart has the reputation of being difficult.[2]
>
> "Eckhart develops this theme with an overabundance of unheard of negation: God is without name, since no one can say or understand anything of Him . . . If I say then, "God is good", it is not true; I am good, but God is not good . . . If I say, further, "God is prudent", it is not certain, I am more prudent than He. If I say, moreover, "God is a being", it is not certain, for He is a being higher than being and a superessential negation. A master says, "If I had a God that I could know, I would not have Him for a God . . . You must love Him such as He is: neither God, nor spirit, nor person, nor image; moreover, one without admixture, pure, luminous . . . (Sermon: Renovamini spiritu mentis vestrae). But in another sermon he says, "When I have said that God is not a star, that He is higher than being, I have not, by that, implied to Him being, on the contrary, I have attributed to Him a Higher being. (Sermon: Quasi Stella Matutina)[3]
>
> "God becomes God when His creatures say: 'God'."[4]

And in another part of his Psychological Talks, he takes up further a theme of the so-called 'Aristotle of Arabiá', Abu Bekr Muhhamad Ibn al'Arabí (which is abbreviated Ibn'Arabí) and he comments:

"The sense and mystical intuition perceive that God is manifested in the forms of all beliefs and that those limitations are necessary, since total knowledge is never in act. But it is in being the servant of 'his' Divine Name in itself, that the mystic is in sympathetic devotion with all the Names."[5]

"The learning of the Gnostic is knowing how to practice fidelity to his Lord properly, that is to say, to the Divine Name of which he is invested in his essential being, but at the same time to listen to the precept of Ibn'Arabí: "Know then in your soul as a material for all the forms of all beliefs. He who has been capable of it, is the 'arif', the initiate, he who sees by God, in God, with the eye of God."

"If the experience of the Prophet has been meditated upon and revived as a prototype of the mystic experience, it is by reason of the exemplariness of the conjunction of Lahot and Nasut in their person. Now, this conjunction is thought of not as a hypostatic union of two natures, (such as the Christology of the Councils), but as a theophanic union, that is to say, that of a Divine Name and of a sensitive form, the appearance in which that name is transparent. The two together, neither one without the other, nor confused one in the other, make up the totality of a Divine Name, one as Lord from that Name (rabb), the other as servant from that Name ('abd); one cohered with the other through the pact of a sovereignty and a vassality, or service of love, making them 'correspondents', one with the other—a pact sealed with the initial act of Divine Love, the Sigh of Sadness with compassion for the nostalgia of the Divine Names calling to the beings which they would reveal. It is then that if one gives up thinking in incarnationist terms familiar to our theology for centuries, but is truly thinking the conditions and structures of the theophanies, it is then that the mystical union appears to us as fulfilling in itself a sympathetic union."

Let us end then with this interjection. Mathesis then, even on the theophanic plane of the sympathetic union is substantially allied with epistemology, as a mutation in the era of knowledge with renewed vision of our own society and of the methodology of research. This does not impede the including of logic as a discipline which fashions instruments to the sciences, as a general platform of the sciences, in which the current progresses of mathematics and its fusion with logic bring to mind the fact that this aspiration expands itself into a practice which is attractive to the scientists. The union of epistemology with mathematics forms a species of 'metascience' we might say, which could be derived through Cosmobiology, for example, in a universal epistemology from all the coincident sciences with that of the sciences called 'human'; it is a metascience which permits the union of sciences and mystical fulfillment.

As Emile Mayerson said: "The epistemological conception based on the analysis of scientific facts and theories drawn from all the disciplines can also qualify as a realistic rationalism."

NOTES

1. Archeometry.
2. Psychological Talks, Mystics and Humanists. XXXIII. 39/5/1.
3. ibid. XXXIII. 40/2/3 to the end.
4. ibid. XXXIII. 41/2/5.
5. Psychological Talks, The Wisdom of Muhammed. XXX 22/end, 11 to the end.

PART IV

LIFE—FORM—THOUGHT

CHAPTER 9

The Supreme Cause Contains The Non-Manifested Effect and the Manifestation

The history of science, says Bachelard, is not able to do without the work of epistemology, as much for the beginning of it as for analyzing its process.

An epistemology on the level of perception (the sensible, the appearances, common experience) and the level of understanding (science, the concepts, the laws, deduction) is syntactic. It includes a metaphysical activity of the dialectical reason, scientific pluralism, and is coherent in that science has the power of grouping and unifying in its bodies of theory.

Epistemology, the middle road between the occupation of the scientist and the traditional philosopher, is going to constitute, in the expression of Hyppolitus, the 'consciousness-of-self' of science, which accompanies it and advances it in every process.

Then the fundamental phenomena of life of the sciences, according to the saying of Francois Guéry, will be the 'epistemological act'.

The true reality in the matter of thought is the epistemological act, which always acts upon the product of something before, which at the end of its work of elimination has produced a new form.

Applied epistemology treats, then, of a coordination of the investigations and their methodology, and on the plane of ideation: to have a kind of examination of conscience at the level of all our culture.

This enveloping concept exalts life to a field most profound, and from the highest level of use of the intelligence, that is, that of investigation, that continually sought-for discovery and revelation of the light of reality, that relation of potencies as broad as the field of the formulation of a problem which is at least more important than the solution. Investigation, the discipline which demands a greater endowment of human genius for going farther from the effects toward the laws and principles, is the cerebral cortex of world thought, and the apex of orientation of criterion in this new age of wisdom.

Therefore, an epistemology of life implies an epistemology of universal applications, an epistemology of existence, that is to say, integral, while at the same time diversified.

Let us say, for example, that for Jean Piaget, who is known also as a child psychology specialist, epistemology is a science which demonstrates the support of scientific learning in a scheme of intellectual and psychological development of the individual. Its reflection arises each time more interior in the sciences, because they produce certain crises or conflicts as a consequence of the internal progress of the deductive constructions, or from the interpretation of the experimental data. It makes it necessary for passing on the hidden or explicit traditions, to propose a retroactive and prospective criticism of the concepts, methods, or principles used until then, in a manner which can determine their own epistemological value; but not as a simple reflection on science, but as an instrument of the scientific process, as an interior organization of the foundations, and above all, in the measure in which it is elaborated by those ones who used those foundations and which is what they need, which reveals incalculable consequences in the future of science.

In mathematics, which particularly interests us in this work, the birth of an interior epistemology is destined to illuminate the general problems of foundation, which is particularly surprising. The "Theory of Foundations", says Piaget, in a field such as this, has ceased constituting then, a 'reflection upon' the science, for deliberately changing in the promoting of technical instruments (to a logical and mathematical time) which forms a body only with the same science.

The methods of Jean Piaget include methods from 'direct analysis' to methods of epistemology which consist in trying to delineate, by simple reflexive analysis, the conditions of thought; 'formalizing analysis' for designating the methods which add to direct analysis of the processes of thought an examination of the coordination between formalization and experience; 'genetical analysis' which endeavors to understand the processes of scientific thought in the function of its development or its own formation. And at this point it is necessary to distinguish between a socio-genesis of thoughts, relative to the historical development within societies and their cultural transmission, and a psychogenesis of the notions and structures which are constituted during the development of the individuals (each time more socialized with the age), as varieties of genetic methods. He determines the historico-critical method, dating from the examination of a present body of doctrines up to the study of their formation, and genetic epistemology, the conditions which try to coordinate the results of fundamental thoughts with the study of formalization.

As an example of direct analysis, comments the same Piaget, we can cite the memorable work of Henri Poincaré in its two aspects: relative to the intuition of number, and relative to geometrical intuition. The general theme of the reflection of Poincaré, as of so many others, was, without a doubt, the adequateness of mathematical deduction to physical reality. In his two fold condition of creator in pure mathematics and of inventor in mathematical physics, he constantly seemed induced to question in what sense the constructions of the subject (in the first of those fields) meet with experience (in the second).

Likewise, the new epistemological horizon which is presented to biology is that of the relations between the organism and the environment; its adaptations to the environment is the field where epistemological biology has to represent, without a doubt, a decisive role in the future.

—

CHAPTER 10

The Consciousness-of-Self of Science

This apex focuses on some aspects of the panorama which is extended toward the future from the application of the epistemology of life which Dr. Raynaud de la Ferrière proposes, and which we have dealt with for the time being only in the sense of its foundations—for we do not want to extend ourselves too much. Let us return to accompany it in what we mentioned as the elevation toward the most remote shores of a new concept of the divinity which does not abandon its causal and scientific support.

"Life", he says, "in its most profound sense, is a realm of precious investigations which ought to be appreciated in its most correct value; the things called sacred are precisely those which have been endowed with the most life."

His concept of the Divinity is composed also of the enunciation: LIFE-FORM-THOUGHT, and he says: "LIFE-FORM-THOUGHT; I have here the point of convergence or final end of all investigations."

At once, he turns this conception to the comparative religions of the civilized world, embracing all things and expressing his own language of union with the fundamental concepts of the religions in the world, in the foundation of a scientific philosophy, capable of identifying the religion. He defines this in "Mysticism in the 20th Century" as 'the intuitive realization of the existence of everything', on the same basis of a scientific criterion, that is to say, in the science, which he defines in a parallel passage in "Mysticism in the 20th Century", as 'the objective demonstration of the existence of everything'. That is to say, the subjective realization founded on the objective demonstration.

It is not precisely the faith but the concepts; therefore, it is not mixed with any analysis.

The platform of faith requires that the concepts of religion be oriented toward a scientific conception which permits the least possible error and the greatest adoption of the constructive principles of each one.

In a certain manner, it is a necessity not only of religion, but of all idealistic platforms, subjective or affective in which man is moved. They are accompanied in a balanced manner by conceptions based and extended by science.

It is the application of discernment. Up to what point did Christianity follow the thought of Jesus? Up to what point did Buddhism follow the thought of Buddha?

History and tradition show us that as soon as religion is withdrawn from the thought of the Great Masters which represent it, it easily falls into fanaticism and dogmatism.

Even better is the exceeding of the religious plane with a universal mysticism, or in the final case, the return of the word religion to its root, religare, to join, to bind, to join man with man, to join man with the Universal Manifestation, to join man with the new concept of the divinity: the Supreme Cause, which contains the effect non-manifested and the manifestation.

> "One must come to prevail over the manner of limiting 'God' with a word (above all when its signification is so poor as the 'God' of the English, 'Gott' for the Germans, 'Dios' for the Spanish, 'Allah' for the Muslims, 'Bhagavan' for the Hindus, etc.) without this signifying to negate the divine existence, as some attempt to do. Likewise, it will succeed in overcoming the adoration of an unknown 'being', for realizing that we are an emanation of the Great All, a part of this great Universal Body: to be unified with this Principle is to achieve intensely the communion with Eternity. Unbounded in our adoration, one wants to say in perfect harmony, it does not matter in what temples, churches, or pagodas, for the ESSENCE-ONE is everywhere always . . . (without special hours). So then, it is not necessary to be

circumscribed to a place of worship, to found new religions, or to have determined hours for the ceremonies . . . Far be it from me to criticize the organized churches. I have always been the defender of order and discipline. Nevertheless, if after such a long time they have not been able to offer an immediate state of beatitude it is necessary to go from there to where one wants, and when one wants.

"One cries that, "Religion is the opiate of the masses". I respond that, "Yes, but to the intoxicated it is not necessary to abolish the drug". I remember also that one day in one of my public conferences an old erudite Spaniard questioned me to say that religions are crutches. That is true, but . . . Why do you want to suppress this help to those who cannot walk alone? Frequently one believes himself capable of walking alone too soon, and the accidents by the atheist are well known so that I do not have to busy myself with commentaries to that respect.

"The diverse channelizations of human thought are indispensable for placing the spirit at the door which leads out to the garden of personal investigation. All the religions have methods of moral conduct it is necessary to respect. Perhaps the only observation that I would be permitted to make is that we have either too many, or too few religions! In effect, 'too many' religions in the sense that all the sects ought to be reduced to a single Religion (from the Latin, religare—to join, to unify, to bind), no matter how identical the bases of all these doctrines which divide the present world are; or, on the contrary, 'too few' religions since we scarcely have 300 presently and because if each individual belongs to a different religion, which would be magnificent, we would be in the presence of a phenomenon of peculiar personal reaction, which would be the special object of psychological investigation. The 300 most important religions and sects that actually exist in the world, are subdivided internally in turn, by reason of the discrepancies of dogma, of details of ritual, or diverse theological interpretations, giving an opportunity to true progressions, which are not filled with pretended religious IMMUTABILITY. There is no immutable religion; in all there have been modifications of their primitive dogmas; in all they have been transforming a part of their teachings. Only YOGHISM remains IMPERTURBABLE through the centuries.

"We would have to say some more on the extreme difficulty of reconciling religion with science, and it is sad to add how difficult it is that Christian religion in particular is explicable by science and that science equally is acceptable by religious dogma. It was first proclaimed, authoritatively, that only Christian dogma would have value in opposition to all scientific explanation; later it was conceded a greater tolerance by religious representatives. One understands now that science as well as religion is in continuous development because neither is unchangeable, both are *subjects* of constant variations. In the earliest times of Christianity it was the general belief among the Christians that the world would have its end at the same time as the people who lived in that age. For a fact, this formed part of the doctrine and discipline in which they were organized, and in the course of time the Christian faith was modified various times in accord with circumstances, since time and events in a certain way thus made religion. More than once the solid scientific enunciations were combatted by the representatives of Christianity.

"In the year 535, the monk Cosmas wrote a book entitled, *Christian Topography*, in which he maintained, basing it directly on biblical texts, that the world was a parallelogram plane whose length was double its width . . . ! At the same time he denied the antipodes. It is necessary to add that it dealt with a traveller who, when he had visited India and Ethiopia, preferred to retire to a monastery in Alexandria, a great center of culture in that age.

"It is recorded that the movement of the Earth was condemned by a Catholic tribunal in the 17th century. In the past century the extension of time located by geological science put the religious populace in a tight spot. The present doctrine of evolution places religion in constant commotion and in this manner science remains incompatible with the spiritual system.

"But it would be unjust to remain believing that religion was always false and science always true. The fact is more complex in reality, and it turns out very hard to explain because in the end it is necessary to make concessions on the two parts. In the course of history, science has not always been right, and religion has not always been mistaken.

"Likewise, the very erudite Jesuit, Petavius, has demonstrated that the theologians of the first centuries of Christianity made use of phrases which, after the Fifth century, would have been condemned as heretical. Cardinal Newman, in a very beautiful polemical treatise, has elaborated on the development of this doctrine, and never retracted it in the course of his life. (He wrote it before being converted into a great Catholic Bishop.)

"Science is still changing more than theology, but this occurs at times because science has not known how to impose with sufficient proofs the veracity of its theories whose conditions religion has argued. At the beginning of the Medieval Age, heaven was in the area of paradise, and the inferno was under the earth; perhaps this definition was not formulated by the highest official spheres of religion, but this order of understanding surpassed by far the standard of a simple people. One will be able to obtain an idea of this kind of explanation in the *Dialogues* of Pope Gregory the Great, a man whose high official position is surpassed by the services rendered to humanity.

"I have used here the word 'religion' in the sense of Christian religion for two reasons: because as I already showed from the beginning of my book (*A Mathesis of Psychology*), it is my desire to direct myself to the Western world which is in particular Christian and, moreover, because I discussed here the fact that I have developed from the Catholic, which I was; that is to say, that my writing is above all the discussion which I have undertaken with myself from childhood, I maintain the controversies in this book as those I have had in thought very often. I have investigated with the methods which were at my disposal and, by those, this work is an evolving continuity of my method, a little like Descartes in his "Discourse". Christian religion is here brought into place because it is necessary to recognize also that no other religion ever ventures on the road of Science.

"In effect, I do not present this book as a method or a treatise but rather it is simply the standard which seems to me has shown the way to follow; it is what as intellectual I have analyzed and as Westerner I have criticized, and little by little, I have opened a way in this apocalyptic labyrinth which is the investigation of the Truth. As Albert the Great said, 'That which penetrates itself, and thus overcomes itself, truly turns toward God'. This is understood, of course, by means of introspection and, by this, that the Buddhist method is perhaps the best for the study of Being and never is in contradiction with true science. (The facts appear as I expose them in *Mysticism in the 20th Century.*). But for my part, I was never able entirely to adhere to this philosophy because it is too passive and I was not able to satisfy my active temperament (a judgment which was completely personal). For the rest no other conversion has taken place in my life. In the course of my studies in search of the Truth, I had not done more than to analyze, scrutinize, and observe the diverse theories which the thinking world offered and whose essence I have tried to extract without binding myself to the dogma which I only studied as documental, following the necessary rites in order to fill me with worship. For this reason I have not abandoned anything. I have not rejected any belief. I have acquired, experimented, agglomerated; they advanced according to my observations, forming thus a synthesis of understandings and revelations as a species of true YOGA in all the meaning of the word.

"Through the arts, science, and philosophy the man searches for ethics, forgetting that an ethics for the whole world does not exist!

"Not even is it possible to find an ethics for all the countries, I say! Nor for a small community, and I refer then to what Ouspensky has so well defined in "Fragments of an Unknown Teaching", page 227: 'One says at times that the European ethics is the Christian ethics, but before all, the Christian ethics authorizes by itself a great number of interpretations, and many crimes have been justified by this Christian ethics! . . . Modern Europe truly does not have anything in common with the Christian ethics in whatever sense it is attributed to it. In any case, if it is the Christian ethics which has led Europe to these terrible wars, would it not be preferable then to remain as separated as possible from such an ethics?' Evidently the whole world always acts 'for the good cause'; and the author continues, page 229: 'Each one is eager to serve the good "as he understands it"! But each one understands it in a different way and consequently men tear and massacre "to serve the good"! The reason nevertheless is the same: their ignorance and the deep sleep in which they live.'

"Ignorance, yes; I say that it is the root of social problems. The lack of understanding produces dogmatism and from this intolerance are born the divisions, the strife, the wars.

"One speaks of Christian ethics, of love, without wisdom, apparently, from the way it is treated. Christian ethics is rather different from what is taught by the thirty different religions which are called Christian. Love is something of which little is spoken without knowing how to give the first steps in this vast realm, and as Master Gurdjieff teaches it well: 'He who wants the Truth will not speak ever of love or Christianity because he knows how far they are from it. Doctrine is for the Christians and these are the ones who live according to Christ, that is to say, who do everything according to his precepts. Can they live in accord with the precepts of Christ, who speak of love and ethics? Naturally, NO! But they will always have charlatans of this kind! Nevertheless, there is a sign which does not deceive: those who speak in such a manner are empty men and they are not worth the trouble to spend time with them.' The teaching of Georges Ivanovitch Gurdjieff is based above all in the work of a group in full identification with the Master."[1]

NOTES

1. A Mathesis of Psychology. pp. 23-24, 147-9.

CHAPTER 11

Feasibility of the Fusion of Religions

Dr. Raynaud de la Ferrière addresses the religions in their own language, or better, in a tactically comparative language on a feasibility of the union of religions for the fusion of them by means of a platform which adapts them to the new concept of the Divinity mentioned before, joined in a scientific-philosophical form: the Supreme Cause contains the effect non-manifested and the manifestation. At the end of all the investigations is the trilogy, LIFE-FORM-THOUGHT.

It is here where it is fitting to indicate that upon focusing at last on the investigations of the trilogy LIFE-FORM-THOUGHT, it tacitly supports Jean Piaget in the trinomial Genesis-Structure-Knowledge. Now Serge Raynaud de la Ferrière raises this trinity to the comparative field, in the proper language of comparative religions, as an application of his thought:

> "This Tri-Unity is constituted by the triangular base of the pyramid (very symbolic), with its apex, it is the first triple manifestation (Life, Form, Thought) explained through all the theological trinities which emanate from a God-Only-Absolute.
>
> "By not being easy for scientific conception, the Trimurti of Hindu theology ought to be, in the first place, well assimilated.
>
> "It invites reflection on a great problem of micro-macrocosmic evolution: BRAHMA is the creator in the sense

of his constructive action, which is manifested by means of the Universal Cause; VISHNU is a guardian because his protective power is manifested in his diverse descents, he is the form which, in all the great periods, comes to be materialized in an archetype; finally, SHIVA is the destructor in the sense of abolition of passions. He is, then, the purifier (symbolized by the Christian Holy Spirit and by the Ideation of our theological theory) emanated from the Absolute (Life-Form-Thought).

"It is trinity which is reproduced in respiration through the three times: aspiration-retention-expiration, and in the three letters of the sacred word A-U-M, it is the Tei-Yang-Yinn of the Chinese, reproduced through the Kether-Hochmah-Binah of the Cabalistic Hebrews, as also the legendary Brahma-Vishnu-Shiva is, and the Sat-Chit-Ananda and the mythological Father-Son-Holy Spirit, whose profound reality corresponds to the Manifested-Christ-Universal Consciousness.

"Kether-Hochmah-Binah, the three first splendors called Supremes, are intellectuals and are the Divine tri-manifestation of the first propulsion.

"To this triunity (Life-Form-Thought) corresponds the Tem-Shu-Tefnut of the Egyptians (in place of the Horus-Isis-Osiris myth) the Creator God, Natural and Manifested in the great mystery of: Life (in place of an anthropomorphic God-the-Father which limits the sincere conceptions) up to the Form, which comes in its regular time to materialize the Divine: as a material emanation from the Essence of Life, as a True Avatar (Messiah): the Christ, as Master Jhesu (in place of the historical Jesus constantly exposed to criticism), as a spiritual realization. Eternal Thought, understood as a state of universal consciousness (in place of a hypothetical Holy Spirit). Thus, passing eternally through these three stages, gives us the example of a circle whose center is everywhere and whose circumference is nowhere.[1]

"All difference is rooted in the fact that some of these trilogies are founded on a healthy reasoning, studied and esoteric, and others in the blind faith of a simple history, without any knowledge and purely exoteric.

"Therefore, we would be within reach of a universal mysticism ruled by scientific thought, in the variables suitable to it; the undulating frontiers of objectivity and subjectivity. The system already expressed of Life-Form-Thought is the emanation of the idea Archetype-Macrocosm-Microcosm symbolized by the ideological trinity Spirit-Soul-Matter.

"The triple manifestation of the Absolute: Life, Form, and Thought is constituted in Christian theology by Father, Son, Holy Spirit, translated in the Hindu philosophy as Sat-Chit-Ananda, without confusing it with the Brahman concept of Brahma-Vishnu-Shiva. The Muslim trilogy El Aquil, El Aqul, El Maqul, is analogous to the sephirotic theory of the Hebrews, where we have Kether-Hochmah-Binah, which corresponds in the Egyptian esotericism to Tem-Shu-Tefnut and which should never be confused in any way with the Great divinities Horus-Isis-Osiris, of that religion.

"The Chinese have Tei-Yang-Yinn, which corresponds as in Yoga to the three forces, Sushumna-Ida-Pingala, which can be associated in different ways.

"It is necessary to admit an Absolute. In the philosophy of ancient India, it was Brahma-Nirguna, and in that of the Hebrews, Ain-Soph (unlimited). Of IT, The Absolute, it ought to be said that 'it is not' in the sense that it is not a 'concrete being', or 'manifested', and not in the sense of what 'does not exist' as a Manifested Principle. Consequently, as soon as God is spoken of, already he is not treated as Absolute (unlimited) but of the God-Creator (limited) of the religions: the Ishwari of India, the Shaddai of the Hebrews, or the Demi-urge of the Gnostics, etc.

"Edouard Arnaud, Professor of the Central School of Paris, has said very precisely, 'Truth pertains to the Absolute and it will always be inexpressible and unknowable. To discover the total enigma of the mystery of the Cosmos in which our planet is found lost, and which is an unrealizable dream for our humanity, it would not be more than to know an aspect of the Truth, which is relative to this grandiose manifestation; it would be to know a work of the Architect, but not the Architect.'[2]

"The two phases of the Great All are identical in essence, and only differ in state and nature. These two phases are: the static Principle called Absolute, and the active Principle called God.

"The one is in repose, the other, in movement. God Non-Manifested, and God Manifested.

"Now it is understood better that the Trinity of the Only Manifested is also the reverse of his own dynamic activity characterized by three movements: It is the Divine Trinity which is found in the basis of all the great religions . . . The first phase of the divine Consciousness-of-Self is the Breath, the Life, the Light (God the Father). On the physical plane it can be given this correspondence in the arousing of the conscience which operates before the arousing of the body faculties.

"Immediately, comes Wisdom, Love, and Harmony: it is the Son. In this, differ the distinct doctrines: that is, in which we see this unawakened attribute, Space-Time, the matrix of a future Cosmos, that is, the heavenly Mother, or what Genesis expresses as, 'Darkness reigned over the face of the abyss.' Other religions see the attribute already awakened (The Water, over which the Divine Spirit was moved) as Genesis mentions it, ruled by the Law of Action, that is, the Son of Divine Thought.

"At any rate; the Son, or Virgin (Mother). There is not much difference of state, being always the second manifestation.

"Then comes Thought in action, the third phase of the awakening of the Active Principle, it is the Ideation: the Holy Spirit.

"In summary, the Father, Son, and Holy Spirit, or if it is preferred, Energy, Form (which is the matter in its manifestation) and the Thought (polarized force), they can be defined by: Spirit, Love, Duty, or Will, Learning, Activity, according to the philosophy or doctrine which is professed . . . but in any case, it seems that, in spite of the apparent difference, they are only differences of terminology, and that, at the base of it, it would be simpler to be understood on the value of the words, in the study of the Great Causes (currents of thought) which impassion the world.

"If it is understood that the world does not begin with this God-Manifested, the superstitious veil is lifted from the decadent religions for glimpsing the profundity of the understanding of the Cosmic Laws which unify the real philosophic principles in the Unity of Eternal Truth."

Serge Raynaud de la Ferrière broaches the comparativity of religions within the concordances of a triunity Life-Form-Thought, and we observe that this facilitates the union of religions as much in the sense of their collective conceptualization as in their individual understanding. It also permits the scientific basing of a union between objective demonstration of the universal existence represented by science and the subjective realization of the Great All represented by a universal mysticism.

Schleiermacher for his part used to say that the universe is an uninterrupted activity and reveals to us in every moment, then produces all form, that it gives all being, through the plenitude of life, a particular existence, which it gives every event from its bosom always rich and profound. It is an action which it exerts on all of us for accepting each thing in particular as a part of the All, each finite thing as expression of the infinite. Nevertheless, we see that the orthodoxy of religions through history has been very remote from expressing and living a reality which only a scientific basing permits as experience of the divine. This reveals the reality of its objective and its causation, such as theoretical satisfaction, that is, the necessity of understanding, or through the situation of man in the world which suggests the practical necessities of interest for man by the deeds of life, and therefore of the hopes and incessant fears which worry him. Feuerbach said, the consciousness which man has is from the infinitude of his being and not from his limitation. Max Miller spoke of the human potential capacity for 'holding the infinite'.

Hegel said that if the word spirit ought to have a sense, such a sense would signify the revelation of itself.

For Durkheim in the case of religion, this is the myth that society makes for itself, in the sense of which society is the reality which mythologies have represented under very different forms, but that it is the objective cause, universal and eternal, of those sui generis sensations of which is made the religious experience—that is, the characters which society considers essential for itself. The totem is, for Durkheim, the society itself, and in truth, in the case of primitive society, he thus takes all his categories for the interpretation of the world. Frazer says that the difference between magic and religion consists in this, that the first has direct control of the natural facts, while the second looks for the ways for propitiating the superior powers which preside over nature.

Turning to mysticism, Bergson has seen it in a dynamic form as a superorganic continuation of the vital élan, the impulse toward a new society founded on the universal love.

Contemporary sociology tends to eliminate the abyss which some interpretations of the primitive world have opened between the primitive mentality and the cultured mentality.

The thought of Dr. Raynaud de la Ferrière has abolished the negative theology which defines God negatively, on the one hand, in the impossibility of coming to him and achieving any communication whatever with the ordinary procedures of human knowing, and on the other hand, in a relation between man and God, that is to say, between God and his creation. Master Raynaud de la Ferrière prefers that "it is not God and the creation, but God as creation." At the same time, he excludes the medieval mysticism which prescinded the way of rational investigation, as it was in Saint Bernard of Clairvaux (12th Century), in whom the defense of the mystical way is accompanied by the polemic against philosophy and of the use of reason in general, with the exception of the philosophical camp of Saint Bonaventure who cultivated equally philosophical speculation and mysticism.

Likewise it distantly surpasses the thought of Kierkegaard who says that the mystic is "he who is chosen likewise in complete isolation, that is, in an isolation from the world and from the human relations which disdains existence, and is degrading to itself, because it is always a degradation to be essentially different from others by a simple accidental, and it degrades God because it makes of Him an idol, and of oneself a favorite in His court."

The final degree, would be more near the concept of Bergson who sees in the dynamic mysticism that which continues the impulse created from life which tends to create forms of life more perfect for man, perfecting the creation of the human species.

The pantheist notion is left behind which was created by J. Toland in 1705 and by his adversary Fay in 1709 whose more mature expression can be looked for in neo-Platonism especially in Plotinus who elaborates in imaginative form the notion of emanation, but only accessible through an exceptional and supernatural flight.

Even better a synthesis is set forth with respect to Hegel for whom "God is God, only when He knows Himself, for his knowing of Himself is His consciousness of self in man, and the knowledge which man has of God progresses to the point of man knowing himself in God."

NOTES

1. The Great Messages, The True Science. p. 285.
2. The Great Messages. 300/3/5.

PART V

SOME OTHER ASPECTS OF HIS THOUGHT ON MATHEMATICS

CHAPTER 12

Anticipating Problems

Modern Mathematics is thoroughly concerned with the study of its fundamentals and the critical study of its axioms, which are not far removed from the mathematical thought of Dr. Raynaud de la Ferrière, that is to say that he, for reaching his different conclusions, was not isolated from critical study, and the criticism was in no way less vigorous.

It is time to look at this criterion in the thought of Dr. Raynaud de la Ferrière for imbuing us with the diverse aspects of his criterion, further raising us with him to other horizons of his vision:

> "We could discuss the penultimate nature of mathematical reasoning, anticipating problems.
>
> "It always disagreeably surprises me to see how unconsciously at times they take the enunciations of the geometrical axioms literally, in spite of knowing that that deductive science is based on logical experience, which is more or less irrational in what concerns the pure truth.
>
> "Certain axioms are only propositions in the process of analysis, such as, 'Things equal to the same thing are equal to each other', but they do not form part of the propositions included in Geometry. Others are indeed typical of this science, as, 'Only one line may pass through two points', or 'A straight line is the shortest distance between two points' and here I would like to pause a moment to point out the erroneous character of

this statement which I have more than once proven incorrect, and I have not been the only one to do this.

"The third geometric axiom has given rise to celebrated controversies. Known as Euclid's Postulate, its proof has been sought in vain throughout history. It is stated as follows: 'From a given point only one straight line can be drawn parallel to another straight line'. The Russian Lobachevsky and the Bulgarian Bulyai have demonstrated that the proof is impossible and Riemann made the analysis in "About the Hypotheses Upon Which Geometry is Founded." Lobachevsky has affirmed that from one point various parallels can be drawn to a given straight line. Moreover, he has enunciated various theorems completely opposed to Euclidian Geometry, but not less logical, and which were never refuted, as in the case of 'It is impossible to construct a figure identical to another with different dimensions', or 'If the circumference of a circle is divided into 'x' equal parts and tangents are drawn through the points of intersection, the tangents will form a polygon if the radius of the circle is small enough, but if the radius is large enough, they will never be similar', etc. . . .

"Reimann's Geometry is not different, in fact, from spherical Geometry, and Lobachevsky's is simply a branch of ordinary Geometry since it is limited to two dimensions, as is that of Beltrami, who, on the other hand, can extend his reasoning even farther, by means of his curves called positives and negatives.

"To interpret non-Euclidian geometries one must first clarify the definitions as is customary in other branches of science.

"Thus: by 'space', we understand that part of space situated above the fundamental plane.

"The 'plane' will be understood as a sphere which cuts the fundamental plane orthogonally.

"The 'line' is a circle which cuts the fundamental plane orthogonally.

"In the same sense we should understand by 'distance between two points' the logarithm of the inharmonic radius of these two points and of the intersection of the fundamental plane with the circle which passes through these two points and which cuts it orthogonally.

"Thus the theorem, 'The sum of the angles in a triangle is less than two right angles', of non-Euclidean Geometry can be translated as, 'If a Euclidian triangle has as its sides arcs of a circle, which produced, would cut the fundamental plane orthogonally, the sum of the angles of this curved triangle will be less than two right angles, and thus no contradiction will be possible."[1]

These examples are sufficient illustration of the critical study of the axioms in his mathematical thought.

NOTES

1. A Mathesis of Psychology, pp. 131-2.

CHAPTER 13

The Vernal Point as the Sensitive Axis of the Equinoctial Precession

It will now be convenient to pass to another aspect of the thought of Dr. Raynaud de la Ferrière in which he discusses Astronomy of Position, part of one of the most important of his investigations, the equinoctial precession and its sociological and geochronological projection. He sets this forth, upon confirming the average value of the course of the vernal point which Plato had given as 72 years per ecliptic degree, in the verifications of the sociological, psychological, and religious changes of the Precessional Eras.

The equinoctial precession opens a field of investigations as much in the historical, sociological, branches as in the ecological. I called attention to this effect in my work, *The Scientific Theory of Cosmo-Biology*, Manual III,[1] from which we will quote a few paragraphs:

"His contribution with the application of the sensitive axis of the vernal point at the intersection of the equator and the ecliptic, the equinoctial precession, lights the way to the sociological, historical, anthropological, archeological, geophysical, and other disciplines.

"The vernal point is a sensitive axis of maximum conductivity as shown by observations of the equatorial electric currents whose effects have reached greater intensity at the equinoxes, during the passage of the sun through the intersection of the equator with the ecliptic, according to observations indicated by the Bulletins of the International Geophysical Year, among others: the Bulletins of the IGY, International Geophysical Year (IGY, Transactions, XL—September,

1959), pp. 273-277; ibid. XXXIX (June, 1958) 582-585; ibid. XXVIII (December 1957) 998-1000, and in the Journal of Geophysical Research, LXIV (May 1959) 489-503.

"An important investigation in Support of the conductivity of the vernal point appeared in the International Geophysical Year in which it was concluded that the equatorial electric current of solar origin arising from the zone E of the ionosphere registers its greatest effects, reaching its maximum intensity, at the equinoxes. Walter Sullivan notes in *Assault on the Unknown* (International Geophysical Year, McGraw-Hill, 1961) that when Laurence J. Cahill Jr., one of the co-workers of James A. Van Allen (discoverer of the famous radiation belts which bear his name, and which are situated completely within the magnetosphere zone located within the exosphere where the ionized particles of cosmic origin remain adhering to it, impeding their passage, as a protective barrier) published the results in 1959, he drew attention to the fact that many questions still remained unanswered on this important subject.

"The pull of gravity causes the inclined axis to recede in its upper part, tracing the surface of a cone.

"The force which causes the Earth's axis to have precession is exercised by the Sun and the Moon on the slight equatorial bulge.

"The effect of this force is to change the direction of the axis, not its inclination: the axis remains at 23°27′ while tracing a cone each 25,920 years.

"The equinoxes are moved with the rotation of the axis, also making a revolution each 25, 920 years.

"Plato had already spoken of this precessional cycle of 25,920 years, with a displacement of 72 years per degree of arc on the ecliptic. Divided into 12 parts, each corresponds to an average of 2,160 years per Precessional era.

"Master Raynaud de la Ferrière verified the average index of this movement, or constant mean of precession, from the point of the observer of the apparent movement in 25,920 years, or 72 years per degree, such as Plato's cycle of 25,920 years, around which the variables of the equinoctial precession, in spite of its insignificance in time, can be investigated precisely for supplying the historic detail.

"For our part we consider that the equinoctial precession is also a movement of orbital ecliptic trajectory and that, whether it is as an 'apparent' movement seen from the plane of the observer, or 'real' from the Earth's axis, expressions stamped on astronomy forgetful of the social function, have variables of its movement which may be taken into consideration to be studied within the average index figures given by Dr. Raynaud de la Ferrière, of 72 years per degree of arc. Said variables, natural to the ecliptic and nutational movement, result, interestingly, in the comparison of historical or collective events, since a minute's difference of arc in the equinoctial precession is equal to more than a sidereal year.

"The curve traced by the Earth's axis is not a smooth circle. It has small waves due to the bobbing of the Earth's axis around the central point of 23°27′. The true movement of the axis is a combination of the precession and bobbing—this last movement being known as nutation (from the Latin word 'nutatio', bobbing). The complete period for one of these waves is 19 years, the bobbing in its highest point is 9 seconds of arc.

"The curve traced by the North Celestial Pole is not a smooth circle. The waves of this curve are nutations.[2]

"For his part, in "Astronomy and Astronautics", the Brazilian Astronomer, P. Jorge O'Grady de Paiva, founding member of the ISRJA (Interplanetary Society of Rio de Janeiro), defines the constant of Precession, taking into account the values both of the nutation of the Earth's axis as well as the proper movement of the solar system in the direction of its apex.[3]

"After transcribing some of Dr. Serge Raynaud de la Ferrière's conclusions regarding the manifestations of the New Aquarian Age, we will give, in the following pages from the "Scientific Theory of Cosmobiology", a comparative picture of some of the characteristics of the Precessional Era that is ending, (Pisces among the Latins began in 1 AD), and of the entering Precessional Era of Aquarius, (Aquarius began in 1948) which is the reason for a broad comparative study on sociological, historical, and religious levels which will be developed in the Acts of the Sciences II, of this theory.

"Let us comment that the course of the vernal point (25,920 years) as sensitive axis, a term coined by Master Raynaud de la Ferrière, permits not only the investigation of the lengthwise passage of the constellations of the ecliptic and their multiple factors, but also its orientation towards each celestial hemisphere, through the ages, in a proportion of 100 turns of equinoctial precession for each 2,592,000 years, approximately.

"These phenomena related with the isothermic changes, in which heat is the principal factor for the physical and thermodynamic phenomena which are produced in the atmosphere, such as the rains, winds, and storms, whose study constituted Actinometry, set forth before the presence of modern cosmobiology, as a comparative actinometry, a biocosmic actinometry of interest for geochronological research, relating some geological eras with an average duration of 2,160 years or with the precessional cycles of 25,920 years.[4]

"It is undoubtedly an investigative necessity to go back to the horizons of history through the Precessional Eras. At length, in the twilight of time, we are led along the road of lost cultures toward geochronology, bearing in mind the climate of the biosphere in the macroscopic cycle of times.

"Let us look, as an Act of Science, at what Frederich Zeuner, Professor of Archeology and Prehistoric Geochronology at the University of London describes in "Geochronology, Dating the Past", which takes its methods from geology, botany, zoology, and physics. Its main object is to form graphs of time, in years, which are extended to a past further back than the historical calendar, which signifies literally a computing of time with relation to the Earth, and implies the attempt to compute in years, different things of stratigraphy, which deals only with relative ages. The term was created by H.S. Williams[5] in 1893 to indicate studies in which the scale of geological time is applied (in absolute time) to the study of the evolution of the Earth and its inhabitants. It encompasses both human prehistory and the geological past.

"Its main field of application is prehistoric archeology and human paleontology, since man's evolution, both from the anthropological and cultural viewpoints cannot be understood with any precision unless the time element is considered.

"There exist, after all, many other fields to which geochronology can be applied, in geology and geophysics, and a chronology of that extensive past of man has been attempted which prehistoric archeology on the one hand, and ethnology on the other, offer us without dates nor written data.[6]

Let us continue with the *Scientific Theory of Cosmo-*

biology: Zeuner's method of solar radiation is based on the cycles of equinoctial precession, obliquity and eccentricity, and covers the Paleolithic and Ice Ages, reaching back over a million years.[7]

"According to the method of his astronomic theory, the problem of the duration of the Ice Age counted in years, has fascinated investigators since the earliest days of Pleistocene Geology. There are two possible means of reaching a solution, and the two have been used many times, to wit: (a) calculating the duration based on the rhythm of sedimentation, and (b) the formation of an astronomical scale of time, starting with the periodic perturbations in the Earth's orbits. The second method is more ambitious since it promises to give more precise figures for the whole Pleistocene Era as well as its subdivisions.[8]

"The contribution to cosmobiological research is compatible with its techniques, both from the vernal point in the plane of the observer, with its 25,920 year duration, and from the heliocentric length of the perihelion. In some cases it is necessary to avoid the old fruitless searches by inadequate calculations as Frederich E. Zeuner himself points out when he says, in the same work, that:

> 'Most early researchers in the field of Astronomy were mistaken, since all set out to describe the cause of the Ice Age and were confident the problems would be solved when they observed that the disturbances seemed to solve the enigma of the Ice Ages and interglacial changes. But they failed in their main objective for the reason explained, and also in explaining the fluctuations of climate during the Pleistocene Age, because of an inadequate combination of obliquity, eccentricity, and equinoctial precession.'[9]

"As Zeuner clearly indicates, it will be necessary to make important changes for future research, since, for example, the duration which A. Penk gives to some more recent glacials can be rectified as well as his extension in latitude since, in the cycles of equinoctial precession, there are stages of heat increase as mentioned previously, and also a greater precession of the extension in latitude of said ice ages. Thus, we must also consider the discoveries of the variables of solar radioemission observed by Luigi G. Lacchia in 1957 from the observatories in Cambridge, Massachusetts, regarding the enrichment of particles in the Van Allen belts, and other variables of radiation and of the equatorial heat, as well as those indicated previously in this chapter concerning the vernal point, set forth in his observation of the changes in historic cycles and of the changes in religious cycles."[10]

This research sets forth in this new field of the future, a new interdisciplinary scientific thought which, with a mathematical basis and controls impossible prior to the advent of computation, permits the study and verification of macroscopic cosmograms in their application to man, to animals, to human relations, to education, to criminology, to socio-historical research and more; on the level of time, as the astronomical theory of geochronology has shown, to the ages of our geoid as well as an actinometry of thermal and meteorological changes, cycles of glacials and thaws, with the corresponding disappearance of prehistoric cultures, according to the cyclic passage of equinoctial precession.

NOTES

1. The Scientific Theory of Cosmobiology. (This was supported rectorally by the Vice President of the 33 Peruvian Universities, Dr. Aníbal Espino Rodriguez, Rector of the National University of Trujillo, Perú)
2. ibid. p. 591.
3. ibid. pp. 600-602.
4. ibid. p. 643.
5. Williams, H.S., 1893 "The Elements of the Geological Time Scales", J. Geol Chicago 1, pp. 283-295.
6. The Scientific Theory of Cosmobiology. pp. 649.
7. ibid. p. 650.
8. ibid. p. 651.
9. ibid. pp. 652-6 and pp. 653/1.
10. The Scientific Theory of Cosmobiology, Dr. D. Ferríz, 1976. Published by the National University of Trujillo, Peru.

CHAPTER 14

Combining and Researching the Representative Value of Numbers

Let us explore now another aspect of the mathematical thought of Dr. Raynaud de la Ferrière:

"Thus it would be possible to combine and scrutinize the representative value of numbers in order not to be lost in a vague symbolism lacking in concrete value; with a little skill it will be easy to remain within the limits of scientific philosophy, providing at the same time mathematical recreations.[1]

"Thus, 'by chance', I had the opportunity to leaf through, on my arrival in Rangoon, a pamphlet of the Buddhist Mathematical Society (No. 1 of 1948), and my attention was immediately caught by the number 58. It says there that the primary cause of the total 207 of mind and matter is found in the 29 types of associated Kama and in the other 29 types of Kama from which pleasure is derived, that is, the total of 58 for both, which is represented by the sum of initial positions occupied by the 9 digits in the value of Pi, for 31 figures, or positions, with a total of 207.

"If we write the formula, 3.141592653897932 38462643383279 (31 figures in all) and we line them up under the first 10 digits in the following manner: the 3 corresponds to the first place, 1 to the second, 4 to the third, and the next 1 to the fourth place, and so on until at last, 2, 7, and 9 are in the 29th, 30th, and 31st places respectively. Then, if we examine the two lines which represent the value of Pi for 31 positions and the way in which the 9 digits

1	2	3	4	5	6	7	8	9	10	11	12	13	14	15	16
3	1	4	1	5	9	2	6	5	3	5	8	9	7	9	3

17	18	19	20	21	22	23	24	25	26	27	28	29	30	31
2	3	8	4	6	2	6	4	3	3	8	3	2	7	9

are placed in it, one can see that the final position of each one of the numbers corresponds to the 31 places as follows:

1 2 3 4 5 6 7 8 9
4 + 29 + 28 + 24 + 11 + 23 + 30 + 27 + 31 = 207

"Now we observe that 58 is the result of the first position of each one of the 9 figures corresponding to the same 31 places:

1 2 3 4 5 6 7 8 9
2 + 7 + 1 + 3 + 5 + 8 + 14 + 12 + 6 = 58

"Here we see the value 58 as a cause and 207 as a collective effect.[2]

"Our attention increases upon considering the number 248 in connection with the value of Pi, according to the President of the Buddhist Mathematical Society.

"The 248 years of the revolution of Pluto constitute the sum of the 164 years of the revolution of Neptune and the 84 years of Uranus.[3]

"We ought to mention the variation 3.141592742 as the value of Pi in the squaring of the circle with a polygon of 12,000 sides. The theories of the mathematician Jaime Udelman in Venezuela, and Professor Henri Varcollier of France vividly interest us, although outside of the philosophical line.[4]

"It is firmly established that only 22 types of matter make up the interrelation between mind and matter, which are presented in the four categories (Will, Mind, Heat-Energy, Fuel-Food). Said affirmations increase the evidence of the geometrical and mathematical base of Buddhist philosophy, according to which six intercepted lineal forces which cross only one circular plane can divide the plane into 22 parts at the most.[5]

"Every graph can be turned into a formula and every algebraic equation can be expressed by means of a solid body or a surface.

"The volumes of more or less eccentric forms are capable also of being made concrete, thus fulfilling the ideas of sages.

"In this specifically mathematical paragraph we are obliged to allude to the great problem of squaring the circle. We consider obvious the well-known solution of a regular poly-

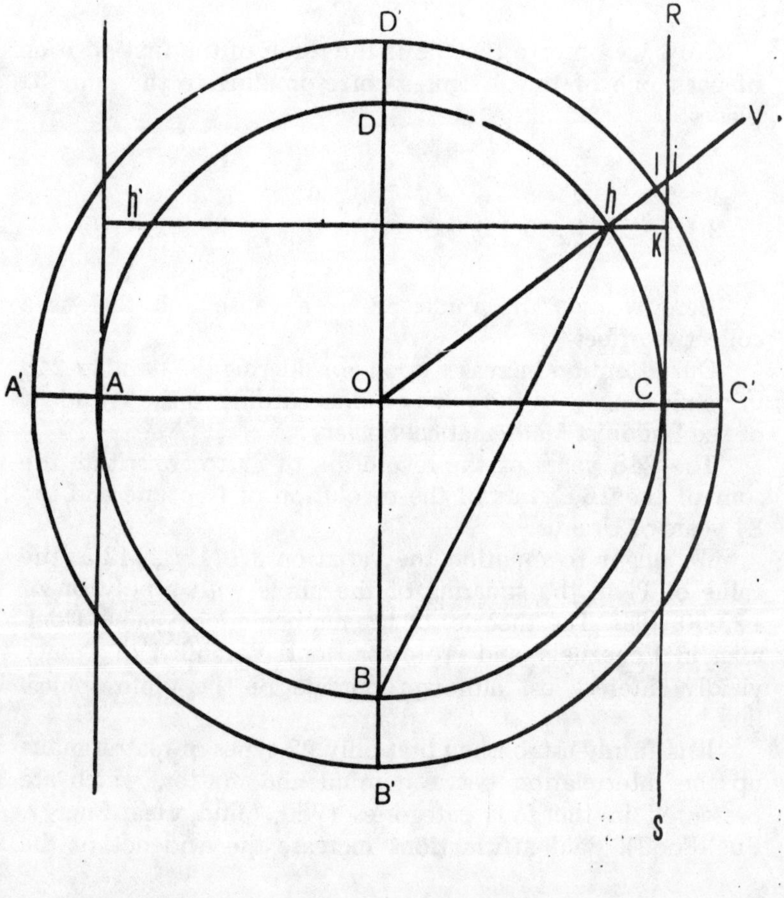

gon of 12,000 sides which approximates almost exactly the value of n = 3.141592742, which is an ideal solution based on geometrical principles.

"Nevertheless, another recent solution is nearer to the truth, discovered by a South American mathematician. A circumference ABCD is drawn with any radius, taken as a conditional unit. On this figure another circumference A′B′C′D′ is drawn concentric, with a radius equal to that of the first circle plus the 10th part of the diameter AC, that is to say, $OC' = AO + \frac{AC}{10}$. Next draw the tangent RS on C and turn the line OC′ with O as its center, until a length is obtained LI, equal to the 10th part of HK. The points L or I are those in which the line OV cuts the circumference and the distance HK is that which exists between the tangent RS. We have thus, the unknown point H which we were looking for to draw the line BH which graphically solves one side of the square of the given circle."[6]

Let us focus upon another brief part of the mathematical judgment of the Sage, Raynaud de la Ferrière, "One knows that the word 'existence', for example, does not have the same meaning when it deals with mathematical things as it does when treating of a material object. John Stuart Mill said, venturing perhaps too far, "All definitions include an axiom." In such matters it is often a question of words, of terminology, or of acceptance of definitions. When in the middle of the past century they laughed at Coulomb's fluids, it was not suspected then that the theory would be accepted 50 years later with the name of electrons. What is the difference between these permanently electrified molecules and the electric molecules of Coulomb? In the electrons of electricity there exists the 'mass' (small particle of matter), but Coulomb never denied 'mass' of his fluids."[7]

NOTES

1. Great Messages 512 and 513.
2. ibid. 505/2a and 506/4.
3. ibid. 512/7 and 513/1.

4. ibid. 510/4.
5. ibid. 510/5.
6. Transcendental Problems, Jaime Udelman.
7. Yug Yoga Yoghism p. 132.

CHAPTER 15

His Integration of Primitive and Modern Mathematical Thought

It is convenient to consider in the study of the living experience that concerns us, Dr. Raynaud de la Ferrière's integration of the primitive mathematical thought with the internal and external factors of modern mathematical thought. This is because the universality of his thought embraces a synthesis in time and his own experience is thus constituted and focused completely within a new age.

In his work, *The Art of Healing*, he makes defense of the profundity of some primitive peoples and states:

> "Modern beings are looking once more at the Old Wisdom, the antecedents from an existence that is more in agreement with the psychological needs of the moment. We have been stating this for almost twenty years and God knows if the stagnant spirits have intended to hinder our work which, like all the works of Precursors, has been called utopian. Be that as it may, official science each day more and more bears witness, in its researches, to the Wisdom of the Ancients.[1]
>
> "We will never know how to sufficiently encourage researchers, including the simply curious, to read the UNESCO *Courier.* In the issue of August 1957, for example, the said bulletin contains articles of the greatest interest for the general culture of men, especially for those who conceal their ignorance, resorting to the word, 'superstition' and who shrug their shoulders

when they deal with questions which exceed their mediocre intelligence."[2]

The following paragraphs by Alfred Metraux appeared in the above-mentioned number of the UNESCO *Courier* in which he says:

> "In virtue of one of those habitual contradictions of the human spirit, the people described as 'primitives' or 'savages' are often praised for their marvelous scientific thoughts. The possession of secret formulas is attributed to them, whose efficacy will exceed that of the products used by our chemists and our physicians. Those persons who, while still scorning these native peoples, speak with admiration of their almost supernatural knowledge, show excessive credulity and candor, but they are nearer the truth than those who deny to native peoples all science worthy of being known."

Alfred Metraux continues, saying:

> "If it is true that the natives of South America do not possess miracle drugs, one should not forget the merit of having discovered the powers of the bark of cascarilla, or cinchona, of coca leaves and the vegetable vines and saps from which curare is extracted. The conquest of the tropics could not have been achieved without quinine, and everyone knows the role performed by cocaine in modern surgery. As for curare, that extraordinary poison which acts on the nervous system, its application is each time more decisive in neurology. These three discoveries are only an example for determining the magnitude of our debt to the so-called primitive peoples.
>
> "The list of scientific conquests of those peoples could fill many pages, if we consider other spheres of knowledge and other geographic regions; but that is not our purpose. One must not forget, however, that it was the 'savages' of the Amazon who showed the world the marvelous properties of rubber and who were the first to use the sap of this plant. For a long time white men continued using the methods of the natives to convert

the sap of the Hevea into rubber. It should not be said that these discoveries were accidental and that no scientific spirit is revealed in them. To extract curare, for example, from the plant *Strychnos toxifera*, needed many and long experiments, and above all, the qualities proper to wise men: a sense of observation, patience, and delight in scientific experimentation.

"The biologists have not always been recognizing the contribution of 'primitive' peoples who have transmitted to us precise and valuable observations."[3]

In another part of his work, we observe that Serge Raynaud de la Ferrière worked with the so-called magic squares, which he points out, "are accepted today by science which needs from time to time to resort to them for finding certain mathematical solutions." And he adds, "From ancient times they have had the quality of solving arithmetical problems and likewise other applications, then more hidden. Magic squares have different functions, of which the main one is, without a doubt, serving as keys of interpretations. It is, in effect, a good screen, where the numbers, placed in a certain manner, can be replaced by letters, and they are likewise valuable in sonometries, and they are able to tabulate various charts of diatonic intervals in ciphered notations.[4]

In the Department of Mathematics at the National University of Trujillo, The Center of Higher Mathematical Studies assembled in December, 1973, with representatives from the universities of the member countries of the Andrés Bello Compact of the Andean Pact, for the University Compromise of Trujillo, the House of Studies, which was founded by Simón Bolívar in Perú, had a fine work by the South American mathematician, Professor Roberto Tejada Rodríguez on "The Magic Squares" and their projection on the 20th Century. (1973) In it he considers the close correlation which exists between the ancient world of esoteric numbers and the modern world of space, and which, in spite of the beauty and depth of mathematical knowledge, have not succeeded in casting out all the mysteries which surround the relations and properties of numbers.

In the early history of numbers, says Tejada, the fascinating pastime of watching the intimate secrets of their behavior was discovered. Numbers were not related to technique nor were they instruments for measuring that obeyed rational laws, rather, they obeyed ethical norms and mystical combinations with exclusively qualitative rank. Curious qualities were found in certain numbers; notable relations were ascertained between numbers apparently unrelated. Strange analogies were used between numbers and geometric figures, and this insatiable search has continued for centuries, until our own times, in which strange things remain undiscovered about relations and properties between numbers.

It is in this world full of mystery, adds Tejada, that the magic squares appear, whose study has stimulated the interest of mathematicians of all ages, such as Fermat, Euler, Bose, etc., and today acquires vital importance, thanks to the studies of the great British statistician, Sir Ronald Fisher.

Magic Squares are considered mathematical compositions consisting of a succession of numbers placed in little boxes of a square, in such a manner that the sum of the numbers found in each row, column, and the diagonal remains constant. The numbers can be of any kind and are not bound by any relation, but are considered normally numbers which constitute an arithmetic progression. Therefore, a chart with side 'n' will contain n^2 boxes, and in these there will be n^2 numbers. Using the first n^2 natural numbers, the constant, or sum of each row, column, and diagonal will be: $\frac{n(n^2+1)}{2}$.

The order of the magic squares is given by the number of rows or columns; moreover the number of the magic squares which may be formed with the natural sequence of numbers, for a square of side 'n' is extremely large as soon as 'n' exceeds 5. For example, the French mathematician Frenicle de Bessy presented to the Academy of Sciences in Paris in 1675, the complete list of 880 different magic squares which may be formed with the first 16 natural numbers.

In the 16th Century, Cornelius Agrippa made up squares for 3, 4, 5, 6, 7, 8, and 9, which he associated with the seven planets known at the time. During the same century a magic square of 4 x 4 engraved on silver protected against the plague and was used pretentiously in psychoanalysis. This was the period when the German painter and mathematician Albert Durero painted a magic square of the order 4 in his famous picture, "Melancholy" in 1514.

This square has even more complex properties than appear in the relations of the rows and columns. The most important characteristics of this square are:

1. The date in which the print was made, 1514, appears in the central boxes of the bottom row.
2. The sum of the numbers of the upper rows is equal to the sum of the numbers of the lower rows, that is, 68.
3. The sum of the squares of the numbers are also equal, that is, 748.
4. The sum of the numbers of the alternate rows (first and third, second and fourth) is also 68; just as the sum of the squares of these numbers also equals 748.
5. The sum of the numbers situated in the diagonals is equal to the sum of the numbers which are not in the diagonals, also 68.
6. The sum of the squares of the numbers situated in the diagonals is also equal to the sum of the squares of the numbers not situated in the diagonals, and also, as in the above cases, is 748.

(Tr.'s Note: The properties are the same for the vertical columns as for the horizontal rows, as they should be in a 'magic square.' JP)

The Square of Four

	16	3	2	13	
68	5	10	11	8	748
	9	6	7	12	
68	4	15	14	1	748

1st and 3rd ⟶ 68 ⟶ 748
2nd and 4th ⟶ 68 ⟶ 749

It is true that in *Art in the New Era*, Dr. Raynaud de la Ferrière, who was also a painter, mentions that picture of Durero. This digression interests us for knowing his esthetic thought where we see that regarding Durero he indicates that "it is obvious that he had other thoughts apart from those of the color and the technique of painting; he is a 'scientist of the canvas,' he sees, and immediately sets out the pattern which is presented but with proportion, and even with geometry at times."

He continues, saying apropos of Durero:

> "He exchanged some works also with Raphael. They may be compared on one ground as a singular and unique thing only, concerning this step forward made by one and the other (with Titian also), as two pioneers which separate fearlessly from the concepts of their period. Durero had an analytical spirit, needing to be read, not only viewed. I can cite only two of his works which demand, without a doubt, a lifetime to analyze and, above all, to understand the sense and thus obtain benefit from them. They are the "Gentleman of the Reformation," (1513), and "Melancholy,' (1514). Without a doubt, it ought to be called "Melancholy" as it is presented on the banner at the base of that picture. In effect, even in the manner of writing the word, Durero chose the symbol to express himself.
>
> "It is the search for the Absolute which guides him, and not art in the esthetic sense; not the manifestation of beauty as Raphael, not even the psychological sounding as in Titian's portraits, but scientifically detailed, at times, in the moral sense, but always, and above all, as a thinker.
>
> "The self-portrait of the artist painted in 1500 has been criticized because Durero intentionally made it a resemblance to Christ. If Durero, at twenty-eight tried to form a personality similar to that of the Great Nazarene, perhaps it was more due to Christ-like ideology than to complacence. In effect, the self-portrait which he painted of himself, in 1493, was for his "Agnes"; and in that age he is presented as an agreeable personality, perhaps with an advantage, but in no way resembling that paint-

ed when he was 28 and which most certainly had mystic aspirations. What does this 'attempted likeness' with Jesus of Nazareth mean? One thing only: the individual's assimilation into the Christ-like conscience, and not as a copy of the personality.

"Many artists today allow their hair and beards to grow; however, they have nothing in common with the resemblance of the popular figure of the Great Master of Nazareth. More than this is required, especially a type of asceticism; the complete physiognomy ought to reflect a special mysticism with the vows of the Nazarene, and especially those who embrace the life of the Essene devotees.

"In reality, a picture by Durero is not to be looked at, but to be read, as I said, to be studied . . ."

It is most important to see the Christ-like concepts of the art of Dr. Raynaud de la Ferrière, and the fact that he has also been a painter. We have not only seen that scientists such as Albert Einstein were artists (Einstein was a violinist), but the relations between art and science show that that conception helps the scientist penetrate abstract notions of a cosmovision of reality. This is an abstraction of the scientific truth accompanied by the impulse of the esthetic sentiment which, apparently recreative, leads us to a mental state accessible to spiritual and esthetic bridges which open new perspectives to our scientific experiences. There are the highest points of coincidence between the proper objectives of art and science.

While art reaches discoveries of universal esthetic values, science, in turn, discovers universal laws. Art follows the way of the experiential abstract to the concrete embodiment with the materials of art, while science leaves the concrete verification for the abstraction of laws. Both are found in the principles and there is one of the points of union in the individuality of the person who participates in both.

Therefore, this was an interesting factor in the life of Serge Raynaud de la Ferrière in the homogenization of the abstract truths of science and of mysticism, raising him to the plane of the Principles and the Supreme Cause.

In the 18th century, according to Tejada, through studies of the great mathematicians Fermat and Euler, a new age began for magic squares which continues until our times. Therefore, for their study, they are separated into 'magic squares' before Euler, and 'Greco-Latin squares' introduced by Leonard Euler.

We continue with Professor Tejada where he says that many methods of constructing magic squares exist, and, moreover, that these methods depend upon the order of the square. Therefore, we will observe three groups: (a) the group of the odd orders, (b) the group of multiple orders of 4, or doubly even, and (c) the group of the simply even orders.

In the magic square of the odd order, according to the method of La Loubre, numbers are always placed consecutively in the square, moving in an oblique direction towards the right (indicated by the arrows) in the following figure. The method, says Tejada, is as follows:

Always begin, placing the number in the center box of the top row. Then go obliquely upward toward the right, and note that the following number, which is 2, appears above and outside the square in the dotted box. In this case, place the 2 in the lowest box of that column. Go now again obliquely up to the right and you will find that the next number, 3, lies outside the square, to the right.

In this case, move the 3 to the box at the extreme left of that row. Immediately following, you will note that the number 1 interferes with the next oblique movement upward to the right; therefore, go one box downward and place the 4 directly below the 3. Now you can manage clearly for the 5 and 6.

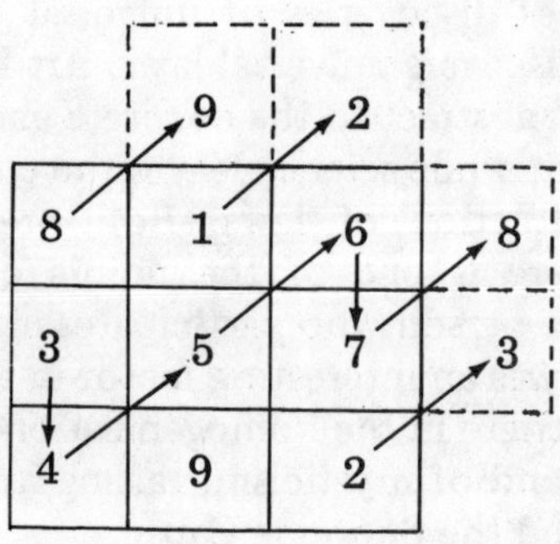

Fig. 1

It is impossible to go obliquely toward the right from this corner, (since there is no box corresponding to this movement); therefore, as before, go down in the square, placing the 7 directly under the 6 and continue as before, placing the 8 and 9 in their proper places, shown by the figure.

The construction of magic squares of double even order, that is, multiples of 4, is based on the construction of the diagonals of each square of order 4; for example, the square of order 4 has any of the following rules:

Begin drawing the two diagonals, and in the upper left corner, begin with 1, without setting it in place for it is on the diagonal. Then, moving to the right, place the numbers 2 and 3 in the next two boxes, but leaving out the 4 from the fourth cell, since it is also on the diagonal. Continue in this manner consecutively, until the forming of the second figure; then place the number 16 (the highest left) in the upper left corner. Count backwards from 16 for each box, and so when you come to the next cell crossed by a diagonal, its number is 13. Continue this process as is shown in Figure 2b, for finally finding the magic square of the fourth order with the first 16 natural numbers. The formation of squares of different orders, following multiples of four, follows a similar process.

The construction of magic squares of simply even order, of the form 2(2m+1), where 'm' is an integer, are quite complex; the Italian mathematician Rafael Leonardi, in 1938, published a method for their formation. For example, for the most difficult of the magic squares of the order 6, the construction would be the following:

	2	3	
5			8
9			12
	14	15	

Fig. 2a

16			13
	11	10	
	7	6	
4			1

Fig. 2b

16	2	3	13
5	11	10	8
9	7	6	12
4	14	15	1

Fig. 2c

8	1	6	26	19	24
3	5	7	21	23	25
4	9	2	22	27	20
35	28	33	17	10	15
30	32	34	12	14	16
31	36	29	13	18	11

Fig. 3a

35	1	6	26	19	24
3	32	7	21	23	25
31	9	2	22	27	20
8	28	33	17	10	15
30	5	34	12	14	16
4	36	29	13	18	11

Fig. 3b

First divide the square of the order 6 into four squares of the order 3, as shown by the dotted lines on Figure 3a.

Begin with the square of the order 3 in the upper left part and form a square using numbers 1-9; then in the square of order 3 in the upper right part, place the numbers 19-27; finally, in the square of order 3 of the lower left part, place the numbers 28-36, obtaining the Figure 3a. So that it will be a perfect magic square, you ought to interchange the three cells indicated by a zig-zag line, i.e., place 8, 5, and 6 where 35, 32, and 31 are, giving as a result 3b.

During the last years of his life, the famous Swiss mathematician Leonard Euler (1707-1783) wrote a voluminous account called *Investigations Into a New Kind of Magic Squares.* In modern times, these new kinds of magic squares are called 'Greco-Latin' squares after the custom Euler had of designating his cells with Latin letters, in opposition to the Greek letters.

Let us consider, for example, the square on the left in Figure 4a. The four Latin letters, a, b, c, and d occupy the 16 boxes which are formed in such a way that each one of them does not appear more than one time in each row or column. Figure 4b represents another Latin square different in that its boxes are identified by the four Greek letters, α, β, γ, and δ, corresponding to the previous Latin letters.

If we superimpose the two squares, as in the right-hand section of the Figure 4c, we verify that each Latin letter is combined once and only once with each Greek letter. When two or more Latin squares are combined in this manner they

a	b	c	d
b		d	c
c	d	a	b
d	c	b	a

Fig. 4a

α	β	ζ	δ
ζ	δ	α	β
δ	ζ	β	α
β	α	δ	ζ

Fig. 4b

a α	b β	c ζ	d δ
b ζ	a δ	d α	c β
c δ	d ζ	a β	b α
d β	c α	b δ	a ζ

Fig. 4c

are called 'orthogonal' squares. The resulting square is called the 'Greco-Latin square.'

In the work, *Introduction to Statistical Analysis*, by Dixon and Massey (1970), there is found an analysis of the intelligence of a group of children to whom different tests are applied. A Latin square of the 4th order is used and the analysis is completed by means of the variations.

Generally, the Latin square of the order 'n' is defined as a square with n boxes per side, whose n^2 boxes are occupied by n different symbols which appear exactly once each on each row and in each column. A combination of two or more Latin squares can exist which are orthogonals by matches.

In Euler's day it was easy to show that there was no possible Greco-Latin square of the order 2. They were known for orders 3, 4, and 5. But, what about 6? Euler formulated the problem thus: Let us consider six different regiments, in each one there exist six officers of different rank. Can those 36 officers be placed in the form of a square in such a way that each row and each column contains one officer from each regiment and one of each rank?

Euler shows that the problem of the n^2 officers, which is identical to the formation of a Greco-Latin square of the order n, can always be solved if n is odd, or if it is 'doubly even,' i.e., divisible by 4. Basing it on numerous tentative fruitless tries, he affirmed, "I don't hesitate in concluding that it is impossible to find a complete square of 36 boxes and the same applies for the squares of the order n = 10, or n = 14, and generally for any order that is oddly even. (For Euler, even, but not divisible by 4.) This proposition is well known as 'Euler's Hypothesis.' It can be expressed more rigorously in the form, "There is no even orthogonal of Latin squares of the order 2 (2m + 1), when m is any positive whole number.

Sir Ronald Fisher (1890-1962), according to Professor Tejada, was ex-professor of genetics at Cambridge University and one of the world's greatest statisticians, and was the first to show about 1920, that the Latin square could be applied to agronomical researches. Let us suppose, for example, that you want to test with a minimum of time and money the effects of seven different chemical products on vegetable growth. How do you organize an experiment which would test simultaneously the seven chemical products and eliminate, at the same time, any bias due to differences in fertility? It is enough to divide the experimental field into parcels which are like the boxes of a seven by seven square, and apply seven treatments with the structure of a Latin square, chosen randomly. Given this structure a simple statistical analysis of the results will eliminate all bias due to variations in the fertility of the soil.

Let us say now that instead of having one variety of plants to subject to those treatments, we have seven. An experiment can be organized to account for this fourth variable (the other three were the fertility in a row, the fertility of a column, and the type of treatment). The Greco-Latin square gives the answer. The Greek letters show where to plant the seven varieties of plants and Latin letters show where to apply the seven chemical treatments. A simple statistical analysis of the results occurs here also.

The Greco-Latin squares are commonly used today in the conception of experimentations in biology, medicine, and sociology, and includes in it market research. The 'parcel' is evidently not a piece of land. It might deal with a cow, a patient, a leaf, a cage containing certain animals, the point in which it is necessary to apply an injection, a period of time, or even an observer or a group of observers. The Greco-Latin square is simply the plane of the experience. Its rows correspond to the first variable, its columns to a second, its Latin letters to a third, and its Greek letters to a fourth. Thus a doctor who practices research can test the effects of five different pills on persons who belong to five different age groups, which presents five groups of different importance who have come to five different stages of the same illness. The most convenient method of experimentation will be based on a Greco-Latin square of the order 5 chosen at random among the diverse possibilities of this order. Additional variables can be taken into account superimposing additional Greco-Latin squares. For a given order 'n,' there is never, however, more than n-1 mutually orthogonal squares.

The story of the investigations which must have conduced Parker, Bose, and Shrikhande to discover Greco-Latin squares of the orders 10, 14, 18, and 22, (and so on, successively) began in 1958 with Parker's discovery which cast great doubt on the exactitude of Euler's Hypothesis. Following the indications given by Parker, Bose established some general rules which would lead to the construction of Greco-Latin squares of a superior order. Bose and Shrikhande next applied those rules and came to construct a Greco-Latin square of the order 22. As 22 is an even number not divisible by 4, Euler's Hypothesis was left invalidated.

When Parker was informed of the results obtained by Bose and Shrikhande, he continued to elaborate a new method which led to the construction of a Greco-Latin square of the order 10, which is represented in figure 5. The symbols of the first Latin square are 0-9, which are represented at the left of each box. The figures on the right are those of the second Latin square. With the help of this square, whose very exist-

00	47	18	76	29	93	85	34	61	52
86	11	57	28	70	39	94	45	02	63
95	80	22	67	38	71	49	56	13	04
59	96	81	33	07	48	72	60	24	15
73	69	90	82	44	17	58	01	35	26
68	74	09	91	83	55	27	12	46	30
37	08	75	19	92	84	66	23	50	41
14	25	36	40	51	62	03	77	88	99
21	32	43	54	65	06	10	89	97	78
42	53	64	05	16	20	31	98	79	87

Fig. 5 Orthogonal Squares of the Order 10.

ence is denied in various works dedicated to experimental methods, statisticians can today, for the first time, organize experiments which take into account four variables of ten values each and control them easily and comfortably.

It is observed that the square of order 3 in the lower right hand corner of the square of order 10, is a Greco-Latin square of order 3. All squares of the order 10 constructed by Parker and his collaborators always contain a square of the order 3, which should be understood as implying a small square can always be formed by permutation of the rows and columns of the large one. By changing the order of its rows and columns, the properties of a Greco-Latin square are evidently not modified. These permutations are, then, of no interest. If a square can be obtained by permutation of the lines and columns of the other, the two squares are considered to be two forms of the 'same' square. At one time it was thought that all the Greco-Latin squares of the order 10 contained smaller squares of the order 3, but this hypothesis was invalidated by many squares which did not present this property.

At that time, the mathematicians concluded in their report. A feverish correspondence was produced between Bose and Shrikhande, on the one hand, and Parker on the other. They made the methods more and more exact; they finally came to show that Euler's Hypothesis is false for all values of n higher than 6. The speed with which they produced a solution for a problem which challenged mathematics for two centuries amazed the authors as much as anyone else. And it is even more surprising to find that the concepts used were still far from the limits of modern higher mathematics.

In the years after 1959, the speed of calculators increased at an enormous rate, while at the same time the skill of mathematicians moved the programming methods to a comparable point. By a method of 'backtracking,' Parker wrote a program for the UNIVAC Military 1206, which permitted him, by dividing a given Latin square, to bring to an end the complete investigation of his orthogonals in a time ranging between 28 and 45 minutes.

With respect to time, the improvement in calculation over the old SWAC was 1 trillion to 1! The consequence was the production of hundreds of Greco-Latin squares of the order 10.

The UNIVAC found orthogonal squares for over half the Latin squares of the order 10 made at random which were presented to it. "Euler, then, was greatly mistaken," wrote Parker, "and the indications obtained from the first calculations show only what this investigation brought to a very broad field."

The great failure of recent work devoted to Greco-Latin squares, concludes Professor Roberto Tejada Rodríguez, rests in not having found groups of three of Latin squares of the order 10 mutually orthogonal. For some time, it has been shown that for a given order 'n,' the maximum number of mutually orthogonal Latin squares would be equal to n-1. A set of n-1 squares of this kind is called a 'complete set.' For

0	1	2	3	4
1	2	3	4	0
2	3	4	0	1
3	4	0	1	2
4	0	1	2	3

0	1	2	3	4
2	3	4	0	1
4	0	1	2	3
1	2	3	4	0
3	4	0	1	2

0	1	2	3	4
3	4	0	1	2
1	2	3	4	0
4	0	1	2	3
2	3	4	0	1

0	1	2	3	4
4	0	1	2	3
3	4	0	1	2
2	3	4	0	1
1	2	3	4	0

Fig. 6: Orthogonal Squares of the Order 5.

the Latin square of the order 2, the complete set only contains then, no more than that one square.

The square of the order 3 has a complete set of two orthogonal squares, and the square of order 4, a complete set of three orthogonal squares. Figure 6 shows a complete set of four Latin squares of order 5 mutually orthogonal. Every pair chosen among those four squares gives, then, a Greco-Latin square by superposition. There exists no complete set of order 6, not even a pair. Complete sets exist for the orders 7, 8, and 9. The order 10 is weaker, and for it there still is not known the possibility of a complete set. It is not even known if there exists a set of three mutually orthogonal squares.

For appreciating the importance of the present study, adds Tejada, we find that Herbert Ryser in his work, *Combinatorial Mathematics* expresses:

> "The interest in the question grows if an account is given of its relation with what are called the 'finite projective planes.' It has been demonstrated that if there exists a complete set of Latin squares of order 'n' mutually orthogonal, it is possible to deduce from that the construction of a finite projective plane of order n, and a complete set can be constructed of Latin squares of order n mutually orthogonal. As Tarry has demonstrated that for the order 6 it was not possible to find even a pair of orthogonal Latin squares, it results that there is no possible finite projective plane of order 6. There exist complete sets (and consequently finite projective planes), for the orders 2, 3, 4, 5, 7, 8, and 9. The order 10 is weaker for there has not yet been proven or invalidated the existence of a projective plane. The discovery of a complete set of nine Latin squares of order 10 would solve all at once one of the major problems which finite projective planes pose. The question is for the moment outside of the possibilities of the calculators and has little possibility of being solved unless a very important increase is produced in the rapidity of those machines or some simplification suggested through a new one."

Professor Tejada finishes his interesting work planning the schemes of apprenticeship, considering the objectives of stimulating the originality and creativity of future mathematicians as well as contributing to the acquisition of the habit of teamwork, encouraging mathematical re-creations.

This acquisition of creativity and mathematical recreation in a classical method agrees with the fact that with modern computer programs, the machine plays in a manner sufficiently sophisticated that its adversary has to be something more than a beginner if he attempts to play chess and defeat it. Such a program was written by Alex Bernstein, M. de V. Roberts, Timothy Arbuckle, and M. A. Belsky of the Service Bureau Corporation. It is designed on the IBM 704, the speed-ordered digital. They say in their article, "Computer vs. Chess Player," published in selections of *Scientific American*, (Editorial Blume, Madrid) that the performance of the terminal is, in itself, a fascinating spectacle. You are seated before the console of the machine, having in front of you a chess table, and you press the start button. After four seconds, the console is illuminated by a panel of lights labeled, 'Delay of Program,' and now you can choose the black or the white. In order to choose the blacks, you must change a commutator of the console. If you decide for the white, it is enough to leave the commutator as it is. Suppose that you have chosen black. To begin the match, turn to press the start button. Now the machine 'thinks' what it will do for its first move. After eight minutes, the terminal prints its move on a sheet of paper.

Let us suppose that the first move of the machine (white) consists in advancing the King's pawn two squares. What it prints, then, is W1. P-K4. The machine continues showing the board with the positions of the pieces, designating its own with the letter 'M,' and those of its adversary with the letter 'O'.

Now it turns on the light, 'Delay of Program,' and the terminal awaits the response of its adversary. You punch your answer on an IBM card, and place this card in a part of the machine which is assigned to read them. For indicating that the move has returned to the terminal, you again press the start button. The machine prints your move and the new position of the board, and proceeds to calculate its second move.

If you have made an illegal move, the ordinator would reject it, printing, "Please, check the last move." At the end of match, after a mate or giving up, the machine prints the score and a message for its adversary saying, "Thank you for the interesting game."

The problem consists in equipping the machine with a system which permits evaluating the merits of the alternatives. If a terminal could give responses prepared beforehand for all possible situations, the chess player would immediately lose his fascination.

The terminal, adds the article, considers businesslike and obstinately square by square: in each square asking if it is occupied. In the affirmative case, it asks whether it is being attacked, or if it is defended, if it can be occupied. All this takes about a tenth of a second, which is a lot if we compare it with the standard of the terminal. Next, the terminal manages to decide which is the best move.

Among the various possible moves (generally, some 30 moves) the machine selects those based on eight questions, set forth by the said authors in the following order:

1. Am I threatened with check? If so, can I capture the piece which gives me check? Protect myself with another piece? Or escape?
2. Are there exchanges possible? If so, can I obtain advantages by accepting the exchange? Or ought I to move my piece?
3. If I have not castled, can I do it now?
4. Can I advance a minor piece?
5. Can I occupy a zone?
6. Do I have some pieces which can be placed on the critical squares created through connected pawns?
7. Can I make a play moving a pawn?
8. Can I make a play moving a piece?

The machine has come to the fourth level, initial play, response, counter-response, and now, the seven potential replies of the adversary to the counter-response.

In all, it examines 2,600 possible positions. It acts as if its opponent were to make the best possible move, within the limits of exploration that the program has been set to.[5]

And the authors comment, further on, that the machine is never distracted, nor does it commit stupid blunders, such as letting a piece be captured 'en passant,' as has happened to all the masters of chess on some occasion. When its adversary is careless enough to expose a piece, the machine instantly avails itself of the opportunity to capture it. In its choice of moves, it often works like a master, making the only move that an expert would consider satisfactory. The machine cannot be classified as a master if its manner of play is considered to the length of a complete game. A machine is not infallible. It is more instructive to contemplate how it loses than to see how to win. After losing a game, the machine returns to repeat happily the same moves and returns to lose exactly the same as before. But they foresee some ideas for programming a machine that avoids the repetition of its errors, and some day—not precisely tomorrow—adds Alex Bernstein and M. de V. Roberts, "we will be able to have machines that improve their games to the measure that they gain experience upon playing against their human adversaries."

In effect, a measure of the complexity of a mathematical problem is given by the structure of its corresponding logical formula.[6]

The human mind can only take in relatively small numbers and quantities. The finite mathematical operations, and the infinite mathematical entities present a notable and fascinating contrast.

Abstract considerations relative to the theory of games can introduce this phenomenon to us in a completely natural form. Dr. S. Raynaud de la Ferrière approaches them with a criterion that helps us in the study of his mathematical experience:

> "The total number of possible cases produces equally probable cases. It is necessary to define the probable by the possible.
>
> "The impossibility of squaring the circle was demonstrated in 1885 and the Academy of Sciences rejected systematically whatever examination referred to this point. There existed a psychological reason in the Aca-

demics in spite of all argument, and if they had been obliged to respond they would have said, "Why do you attempt to expect that a particular value of a transcendental function is an algebraic number if Pi is the root of an algebraic equation? Why do you think that this root is a period of the function sin2x? Why does not the same thing occur with other roots of the same equation?

"If we take the 10,000 logarithms of a table, it is easy to suppose that in the midst of them exists the probability of finding even numbers in the third decimal, and in effect, consulting said tables we find such numbers, even as uneven.

"Taking into account the probability in the physical sciences, we can stop an instant in the calculation of probabilities applied to games of chance.

"The question of luck in roulette has the same foundation as the current calculations in the mathematical hypothesis. The roulette of the casinos is divided into 37 equal parts on which a ball bounces owing to the spin given to the roulette wheel, and the ball is going to stop, after a determined number of turns, in one of the boxes marked with a number (0 to 36). These subdivisions are colored red and black and the choice of these will correspond half to red, and half to black.

"According to the force of the impulse given to the ball, the probability of the angle can vary from θ to $d\theta$. The celebrated mathematician Henri Poincaré supposed the probability from $\phi\ (\theta) d\theta$. He says that the chance becomes completely arbitrary. If we take a ϵ as measure of each red and black compartment and we calculate the integral: (1) $\phi\ (\theta) d\theta$, applied thus to all the reds and to all the black compartments, the results can be compared. It is necessary to consider an interval 2ϵ which includes two compartments red and black consecutively. M and m represent the maximum and minimum with respect to the values of the functions $\phi\ (\theta)$ in said interval. The interval applied to the reds will be smaller than $\Sigma\ M\epsilon'$ and the interval applied to the black compartments will be larger than $\Sigma\ m\epsilon$. The difference will be then less than $\Sigma\ (M-m)\epsilon$.

"Nevertheless, if the function ϕ is supposed continuous and if, on the other hand, the interval ϵ is reduced with respect to the sum of the angle described by the pointer, the difference M-m will be very small. The difference of the two integrals will thus be very reduced and the probability will be about ½.

"Professor Poincaré insists on the following fact: 'We see that even without knowing anything of the function ϕ we ought to act as if the probability was ½.' This explains why, from an objective point of view, if we observe a determined number of moves, the results of these will always be as many reds as blacks. In spite of this logic, it is a perfect error (and the players always fall into it) to imagine, for example, that after a series of six red plays, they play a 'sure hit' on the black color, believing they follow a law that will give them a win because of the difficulty of the same color being repeated seven times. But in reality, they forget that the probability ALWAYS remains ½. Therefore, if the series of seven reds is a difficult result it is no less difficult that a series of six reds is also followed by a black 'hit.' That occurs to them simply because it is easier to notice the improbability of seven consecutive hits on the red than it is to see the improbability of six reds followed by a black.

"The majority of people are afraid to think, to reason. They live from data loaned from others. As Einstein says: 'He would be badly received who would dare to criticize the official teaching which he has received'. This sage writes in his first part of *The Theory of Relativity*: What necessity is there however for the assertion that these propositions of Euclidian Geometry are precise? Geometry depends on determinate concepts like the 'plane,' the 'point' the 'right line,' with which we are able more or less to associate ideas that are defined by certain axioms, which, in virtue of those ideas, we are disposed to accept as true.

"The word TRUTHFUL ought not to be associated with pure geometry because by 'truthful' is understood the designation of a correlation with the REAL. Geometry is not connected with those relations of ideas but only with the logical contact of those ideas among themselves.[7]

"Albert Einstein defines well this relativity of the True and Reality, seen from certain points, when he says: I put myself in the small window of a railway compartment, and let fall upon the embankment a rock, without throwing it. I then see this rock fall in a right line to its rest. A pedestrian who observes from along the track would see this rock fall to the ground describing a parabolic curve. I ask myself now, "Is it that the trajectory followed by the rock follows in reality a right line or describes a parabola?"

"From this simple matter, nevertheless, an enormous problem is raised referring to space, of which we have an idea about its structure, and we ought to look at it with respect to its motion with relation to a rigid body which serves us as a reference. This position now plays a role in the question of Professor Einstein: Which ought we to take as a point of relation, the carriage or the embankment of the railway? Here is where the system of coordination occupies a place. And the celebrated sage concludes, 'The rock travels in a right line relative to a system of coordination rigidly subject to the railway carriage, but in relation to a system of coordination rigidly subject to the ground (the embankment of the road), it describes a parabola.'

"There is still something to add with reference to the fourth dimension. And, as Albert Einstein writes in *Relativity; The General Theory*, Chapter XVII; The non-mathematician feels dominated by a mysterious shudder when he hears talk of the fourth dimension. And, yet, there is no better exposition of good sense than that the world which goes around us is in a space-time continuum of four dimensions. Space is a continuum of three dimensions. That is to say it is possible to describe the position of a point by three numbers (coordinates) x, y, z, and that there exists also an undefined number of points in the vicinity of that one whose position can be described by coordinates as x′, y′, z′. These can be as close as we choose to the respective values of the coordinates x, y, z, of the first point. It is in virtue of this last property that we speak of the continuum and thanks to the fact that three coordinates exist, we can express ourselves as if there were three dimensions.[8]

"In the same manner, the word 'world' employed by Minkowsky expresses four dimensions because it is the world of the physical phenomena in the sense of space-time.

"It is a fact that, according to classical mechanics, time is absolute, that is to say independent of positions and conditions in the movement of a system of coordination. This was expressed in the last equation of the transformation of Galileo: $t' = t$.

"The consideration of the world from the point of view of the fourth dimension is very natural for the theory of relativity, though according to this theory time lacks independence."

Upon approaching the living experience, (*vivencia*), "I realized God through Mathematics," we have considered an order which permits the following of it not only logistically, but experientially. Perhaps some of the first questions that will arise will be the following:

How does this living experience arise, and in what way is it to be lived, and what are some alternatives for following it?

Does it belong only to the mathematical field, or does it involve scientific and social thought also?

What did Dr. Serge Raynaud de la Ferrière think about numerical science which is surrounded by the mathematical environment, and what were his goals?

What mathematical and sociological vision did he have?

What was his transcendence towards sacred thought?

These and similar questions may arise, for which we present now a series of the first alternatives for following this living experience.

The first alternative which we propose is through the way: Mathematical models—biological structures—structures of knowledge—as a logistical realization.

NOTES

1. Psychological Talks. VIII. p. 88.
2. ibid. VIII. p. 81.
3. ibid. VIII. p. 79.
4. Black Book of Freemasonry. p. 92.
5. Computers and Computation, Ed. Blume. pp. 123-4.
6. ibid. p. 158.
7. A Mathesis of Psychology, by Dr. S.R. de la Ferrière. pp. 133-5.
8. Albert Einstein, Relativity. Chapter XVII. pp. 55-6.

PART VI

THE FIRST ALTERNATIVE FOR FULFILLING THE LIVING EXPERIENCE: "I Realized God Through Mathematics"

CHAPTER 16

The Way: Mathematical Models—Biological Structures—Structures of Knowledge; Relation of the Individual With Society—Nature—Universe—Supreme Cause

This living experience which points out new standards to modern spirituality is an archetypical thought of a new age.

If Rachevsky presents a projection of mathematical biology, the thinker Dr. Serge Raynaud de la Ferrière focuses on the epistemological basis of a mathematical ontology. This universality of understanding harmonizes with the glimpses of super-realities that some physicists and mathematicians fathomed.

Likewise, we give an explanation of this experience in examining the way of its discovery in order to place it within reach of scientific thought and individual fulfillment. We would be able to place ourselves, for example, in the process of mathematical abstraction that, through its models, arrives at the biological structures which at the same time carry us to the genetic analysis of the structures of knowledge. This is the same thought in which arises the psychological problems of acquisition of the knowledge which leads us to an epistemology of the relations of knowledge between the subject and the objects; between the individual and the universe; between the person and the manifestation of the Great All; between individual co-existence and universal existence. In other words, the living experience from the mathematician to the Supreme Cause.

That is to say, in the presence of such a living experience, how would it follow from that? In beginning a universal penetration into the study of the thought of Dr. Serge Raynaud de la Ferrière, we would be able to show them this first way as a didactic possibility that permits us to unite ourselves by some means to it. Let us imagine then, the parallelism of four structures.

Let's form the idea of the mathematical model of the Tetract of Pythagoras, whose discovery was considered so important that Jamblico invokes it in his couplet on doric form in which he finds the root of the origin of nature. It was the succession of the first four numbers, 1, 2, 3, 4, considered as successive and joined:

$$(1 + 2 + 3 + 4 = 10)$$

by means of the formula which Dr. Raynaud de la Ferrière calls the Secret Value, inasmuch as the couplet cited by Jamblico also points out that it deals with the oath not to divulge the secrets of that order, and which is:

$$\frac{n(n + 1)}{2}$$

Based on it, we see the model:

$$(1 + 2 + 3 + 4 + 5 + 6 = 21)$$

In it, there is an accumulative sequence whose structure of succession and joining equals 21.

As a structure we are able to bring it, with its parallelism, to the visualization of the biological structure of the polypeptide chain in growth:

(MET–GLN–ARG–TYR–GLU–LEU)

(methione-glycine-argenine-tyrosine-glutamine-leucine)
(1 + 2 + 3 + 4 + 5 + 6)

which, when it is complete, constitutes the protein whose message was originally coded as *DNA*.

But at the same time we can find a parallelism of the same order in the field of the structures of knowledge in the case of the primary school learning of some countries of the world.

(grade 1 + 2 + 3 + 4 + 5 + 6)

After six years of primary study one attains a content (of knowledge) which would come to be equal to his structure of knowledge, that is to say, the total contained in his progressive grades, which we could cite as:

$$(1 + 2 + 3 + 4 + 5 + 6 = 21)$$

If we carry this structure comparatively in the correlative mental process to the field of the sequence, society-nature-universe, we might be able to have the following experiences:

1. subject - object
2. subject - society
3. subject - nature
4. subject - universe
5. subject - universal laws and principles
6. subject - Supreme Cause

We set forth this mathematical way to the Supreme Cause that is able to be purely scientific if it is realized through the experience of the mathematician (mathematical models) who knows biology (biological structures), and who comes to know psychology (genetic analysis of the structures of knowledge) as well as penetrating scientific thought (science in the unlimited sense of knowing, including researchers sufficiently aware interdisciplinarily of these specialties, with a view also to the light of epistemology).

Obviously if we cause to intervene a notion in which gnoseology is included with a blend of philosophical character, then perhaps the unifying criterion of the genetic analysis of the structures of knowledge will be sufficient, that is, the mathematical-psychological way, without overlooking the biological structures. Yet, the analysis of scientific character proceeds more clearly inside the alternative: mathematical models—biological structures—analysis of the structures of knowledge—relations of the individual with the object, in the investigation of: law—principle—Supreme Cause.

When the general theory of the logico-mathematical structures provides the biologist sufficiently precise models (geometrical, cosmobiological, or algebraical) as Jean Piaget might say, for deciding about the reducible character of the organic structures, a great step is taken in the genetic analysis of the structures of knowledge.

In their turn psychology and the human experimental sciences offer us epistemological problems which continue to those of biology, (whose proper field of study is human activity) and of necessity ends by questioning the diverse possible ways of acquiring knowledge by the subject in general—be that subject child or adult, be it the individual or the society, etc. They are the relations of knowledge between the subject and the objects. It is clear that these relations present, at one time or another, in addition to the logical questions of validity, feasible problems relative to the subject. On the one hand, the object is only known by experience, and on the other hand, the experience is always that of a subject. It becomes indispensable then, to know how this is organized, as in the case of genetic epistemology the study passes from the stages of minimal knowledge to the stages of more rigorous knowledge, since the problem of the growth of knowledge effectively constitutes its specific problem.

A little in the manner in which Piaget concludes his genetic epistemology, the interdependent connection of direct and genetic analyses in this explanatory process concerns the fundamental necessity of a dialectic of genesis and structure, corresponding to its alternative interactions. This is because the origin does not exist without structures, since all origins consist of the progressive transformation of a former structure under the influence of new situations. Also, all origins flow into the construction of a new structure, and, as Piaget adds, all genesis, even when it begins and grows under the character of partial imbalances, sooner or later consists of a reestablishment of a new form of balance, which corresponds to a new structure. In this case: mathematics—biology—psychology—investigation—law—principle—Supreme Cause.

CHAPTER 17

Is This Logistic Realization Possible Without the Former Way?

Up to what point does mathematical experience, with its action, creativity, and structural abstraction, permit bridges of ideation that lead the mathematician to a logistic realization of the existence of the Great All, without the necessity of the way which we have just seen through biology and genetic psychology?

Let us remember that Walter R. Fuchs, in *The Electronic Brains*, says that mathematics as activity is a natural thought in which the mathematician 'makes something,' 'works' with symbols, 'constructs' figurations with signs, etc. This discovery of 'making' of action itself as such, makes of the truth a direct vision of dynamism in which all being consists.

"That being can be the individual, but also the universe, as soon as being is conceived as a set of laws and facts associated with attempts of interpretation in its methods and vision of a whole, much in the manner of Einstein in which the universe is full, in the measure in which it can be, since it is balanced but static. It is like the universe of the Dutch astronomer, De Sitter, who, in 1917 considered it comparatively empty while having the appearance of flight, or in the manner of the Abbot Lemaitre who also considers a theory of the expanding universe in which the universe starting from a compact state (the primitive atom) has been dilated, passed slowly through the phase of unstable equilibrium of Einstein, and then continues its expansion. They are different interrelated models, but are always sustained by the ideation of observation and mathematical supervision, being based on a cosmovision, of mathematical models and action.

I have here other points of ideation for the logistic realization of universal existence.

Action can signify, according to Blondel, in *L'Action*, 1936, the result obtained, the work achieved, the end realized, or a species of living creation where the efficacy and the finality have succeeded in uniting, valuing all the mediating powers which have served for that marvelous innovation, called forth by that small word so full of riches: work. Therefore we see what Lutzen Bronwer, Dutch mathematician, said, that "Mathematics is more an action than a teaching."

The notion of action brings us the sense of movement and Norbert Wiener, through his theoretical investigations, succeeded in verifying the hypothesis of Perrin on an 'idealized' Brownian movement: the movement of the particles which could be described as a statistical set of curves, that is to say, the 'idealization' of a movement.

The notion of action extended to the space-time continuum and the fourth dimension, if we could resort to the direct source, is contained in this explanation from Dr. Raynaud de la Ferrière, in *A Mathesis of Psychology*.

> "There is still something to add with reference to the 'fourth dimension,' and as Albert Einstein describes in *Relativity: The General Theory*, Chapter XVII, The non-mathematician is seized by a mysterious shuddering when he hears of 'four-dimensional' things, by a feeling not unlike that awakened by thoughts of the occult. And yet there is no more common-place statement than that the world in which we live is a four-dimensional, space-time continuum. Space is a three-dimensional continuum. By this we mean that it is possible to describe the position of a point (at rest) by means of three numbers (co-ordinates) x, y, z and that there is an indefinite number of points in the neighborhood of this one, the position of which can be described by coordinates such as x_1, y_1, z_1, which may be as near as we choose to the respective values of the coordinates x, y,

z, of the first point. In virtue of the latter property we speak of a 'continuum,' and owing to the fact that there are three coordinates we speak of it as being 'three dimensional.'

"Similarly, the world of physical phenomena which was briefly called 'world' by Minkowski is naturally four-dimensional in the space-time sense.

"It is a fact that, according to classical mechanics, time is absolute, that is to say, independent of the position and conditions of movement of the coordinate system. This was expressed in the last equation of the Galilean transformation: t′ = t. The four-dimensional mode of consideration of the 'world' is natural on the theory of relativity, since according to this theory time is robbed of its independence. This is shown by the fourth equation of the Lorentz transformation:[1]

$$t' = \frac{t - \frac{v}{c^2}x}{\sqrt{1 - \frac{v^2}{c^2}}}$$

Elsewhere in *A Mathesis of Psychology*, he talks about the succession of the continuum:

"Starting from the theory that the sum of the angles of a triangle are less than two right angles, I put myself in contradiction with the customary official current which considers mathematical argument as truly deductive, geometry as derived from experience and physics as based in the induction obtained from the repetition of experimental phenomenon.

"One knows in the manner that Leibnitz wanted to show, two and two are four: Take the number one as being defined, and thus perform the operation "by one more," that is to say, add unity to a number 'x.' This definition does not enter the framework of reasoning. At once, I define the numbers 2, 3, and 4, by the equalities: 1 + 1 = 2; 2 + 1 = 3; 3 + 1 = 4; and in the same manner I define the operation, 'x + 2,' by the relation, x + 2 = (x + 1) + 1. And, 2 + 2 = (2 + 1) + 1, by the definition of 4. (2 + 1) + 1 = 3 + 1, by the definition of

2. And, 3 + 1 = 4, by the definition of 3. Then 2 + 2 = 4, which is what it was required to demonstrate.

"In reality, this is not a demonstration; in mathematics this is called a verification. The argument of Leibnitz is purely analytical, since verification differs from demonstration in that it is the analysis which predominates in it. It uses conventional definitions. If mathematics thus found itself reduced to a series of verifications, it would not make up a science, just as capturing a piece in chess doesn't originate another science either.

"So, we could discuss about the nature of mathematical reasoning, astounding the geometer in his work, checking the various branches of science, but these studies are in constant evolution and I prefer to say as Henri Poincaré in the last phrase of his book, *Science and Hypothesis*, 'I avoid making a prophecy which could turn out false, between the day on which a book is ready for printing and the day on which it is placed before the public.'

"Once cleared of the current calculations, it is the continuum which includes everything of interest for investigation, which Monsieur Tannery in *Introduction to the Theory of Functions of One Variable* has analyzed very well. The continuum is, in effect, unity and multiplicity, but the true mathematical continuum is very different from that of physics and of metaphysics. The elements of the continuum would be intimately bound up forming an all in which the point would not have previous existence on the line, while the line exists preceding the point . . .

"One knows that between two consecutive series can be intercalated one or various intermediate series and continuing, between those rows, others and even more, and so on indefinitely. We have then an unlimited number of numbers of terms and these are called, fractions, rationals, or commensurables. But between these terms which are already infinite in number, other terms can be intercalated, which are called irrationals or incommensurables. The continuum conceived in this manner is not already a collection of individuals arranged in a certain order (and infinite in number), but in external relation one with the other.

"This is not the current concept, observes Monsieur Henri Poincaré, who adds that, according to this concept, it is supposed that there exists an intimate connection between the elements of the continuum, constituting an all in which the point does not have existence preceding the line, but that the line is that which has existence prior to the point. It is a pure creation of the spirit without a true part from experimentation, by the fact that the mathematical continuum is constructed on a continuous scale of irrational numbers and fractions using strictly the whole numbers, as Kronecker and the mathematicians of the German school have done it.

"It is necessary to pass on the definition of the incommensurables, on that which is the physical continuum, and the creation of the mathematical continuum and so many other branches of analysis that they cannot be discussed here, and for which a broad preliminary exposition is necessary before truly bringing up the problem.

"The majority of people are frightened of thinking, of reasoning, and live by facts given by others. As Einstein says: "He would be greeted badly who will dare to criticize the official teaching, which he has received."[2]

After this reference which introduces us into his mathematical vision, let us see a little of the signification and achievements of mathematics for our objective of leading these possible living ways.

The thought of Einstein himself, in his superconsciential vision of mathematics, reminds us what the necessity of thinking has done, that philosophy has always considered mathematics as one of the principal objects of its investigations. These have been brought at last within logic, within the theory of understanding and of metaphysics, until a special discipline has been made, philosophy of mathematics, which has had for a mission to formulate with the greatest possible clarity the basic problems (the 'fundamental problems') concerning the said science.

We will observe only that the idea—predominant for a long time—according to which mathematics is the science of quantity, cannot be maintained. Indeed, there are mathematical disciplines, like topology, which are not concerned with quantity. For this reason, it has been attempted to find a more general concept for defining the content of mathematics: the concept of order.

Mathematical entities are obtained by means of abstractions effected by starting from experience; they are, by thus saying it, highest idealizations of our essential perceptions.

For intuitionism, defended among others by L.E. J. Brouwer and Ared Heyting, mathematical entities can be spoken of only if we can mentally construct them. Intuitionism seems determined to curtail a good part of its mathematical theories of constructions, above all, when it adopts a doctrine which affirms that there are only mathematical entities when they are mentally constructed.

They would come to favor the feasibility of which the living experience, "I realized God through Mathematics," makes up a direct realization without passing through the genetic process of biological structures.

Mathematics can be applied to reality because this is mathematical nature. The mathematical nature is of the true.

Nevertheless, the direct way does not exclude the necessity of the course set out previously through biology and genetic psychology as the form of a process purely scientific and correlated, without which, on the contrary, it is a more logistical and more amazing way through the other sciences thanks to the properties of numbers. Therefore the most plausible in the problem of the relation between mathematics and science—or the sciences—is to adopt a point of view that mathematics is definable in some manner as a language, and that, consequently, its relation with other sciences consists, at last, in the relation which exists, or can exist, between mathematical language and that of other sciences.

According to Aristotle (Met. A 5, 985b23-986b8), the Pythagoreans supposed that the elements of the numbers were the essence of all things, and that the heavens were harmony and number. Upon combining the numbers, their properties turn out so surprising that the Pythagoreans looked for analogies everywhere among numbers and things and came to establish a group of mystical numbers with enormous influence on all the ancient world.

The numbers were considered, moreover, as principles. As

Aristotle says in the same passage cited before, they had within the Pythagorean School a faction which affirmed the existence of ten principles or fundamental oppositions, corresponding to the 10 first natural numbers:

1st:	Limited	—	Unlimited
2nd:	Even	—	Uneven
3rd:	One	—	Many
4th:	Right	—	Left
5th:	Masculine	—	Feminine
6th:	At Rest	—	In Motion
7th:	Straight	—	Curved
8th:	Light	—	Darkness
9th:	Good	—	Evil
10th:	Square	—	Oblong (rectangular oblong)

It deals with a table in which one can see a moral signification: the first terms represent, in effect, the perfect; the second ones, the imperfect.

In this way it can be understood how it is possible to establish analogies and therefore, global unity among concepts whose significations are very different, such as, for example, between the limited and the light and the unlimited and the dark. In every case the harmony corresponds as much to the plane of nature as to the moral order. The relation between the cosmic order and the social order constitutes a very influential aspect from the Pythagorean thought.

In the primordial thought of Serge Raynaud de la Ferrière which we are studying in this book there coincides also his relation between the cosmic order and the social order with the beginning which he sets forth by the astronomical equinoctial precessions of a new Age of true investigations in which the world can be united by wisdom. It is a little like the stage of mathematical activity.

The history of mathematics perhaps began with the discovery that man has in his own body, the full module for measuring and counting, (elbow, foot, arm, etc.). The hands were (and still are among primitive peoples and children) the first abacus. The written Roman numeration recorded it indelibly: the I is the finger, the V is the hand extended, the X the two hands opposed. By this the number 10 is the earliest numbering base of the majority of peoples. The Chaldeans, for practical reasons, evolved toward 60 and left us the happy division of the circumference in 360 degrees. It is not easy to know the centuries or millennia that men delayed in reaching these first rudimentary phases of numbering, until achieving a clear vision of what is called notation by position, that is, of the relative value of the units according to their position. The great moments of mathematical activity correspond historically and experientially with the phases of invention or of renovation of Oriental peoples, (for example Chaldeans, Egyptians and Hindus), in the emotional moment of the discovery of the number and of its notations, with a practical intuition not free of magical and mystical attributes, that still reappear in Pythagoras and Plato, in which truly notable results are obtained.

This brings us to consider, at present, a way not exclusive of the scientific process that crosses through other sciences for reaching the thought of "I realized God through Mathematics," but that serves as a complementary bridge. It is neither indispensable nor less than scientifically requisite that the direct way is a bridge of identification with human thought, that is to say, the fundamental conception of mathematics which considers it as the science that has for its object the possibility of construction—mathematics as construction of concepts called commonly Intuitionism. According to Brouwer, one of the principle representatives of Intuitionism, mathematics is identified with the exact part of human thought, bringing out an intuition which permits presenting the evidence of the concepts and the conclusions. The conclusions, therefore, should not be derived necessarily in virtue of fixed rules contained in a formalized system, but the conclusion can be directly controlled by its own evidence . . .

Intuitionism, upon defining mathematics as the science of possible constructions, does not appeal (as Kant did), to an *a priori* intuition of space. Keyting has thus resumed the point of view of Brouwer: 1) Pure Mathematics is a free creation of the spirit, and does not have in itself some relation with the facts of experience; 2) Simple verification of a fact of experience always contains the identification of a mathematical system; 3) The method of the Science of Nature consists in joining the mathematical systems, contained in the isolated experiences, into a purely mathematical system constructed with this purpose.[3]

Nevertheless, we could not leave off considering, in spite of the inclusion of Intuitionism, the profound and operative relation of our mentioned process: the process mathematical models—biological structures—structures of knowledge—relations of the subject with the object—law—principle—Supreme Cause. Likewise applied mathematics uses the results of pure mathematics for expressing and relating the dependence between objects and real phenomena, deducing beneficial experimental consequences for the experience, "I realized God through Mathematics." Therefore in all natural and technical sciences, they constitute thematics, the principal base of their operative methods. (Physics, Chemistry, Astronomy, Cosmobiology, Topography, Statistics, Insurance, Accounting, etc.)

NOTES

1. Albert Einstein, Relativity, the Special and General Theory. pp. 59-60.
2. A Mathesis of Psychology. pp. 130-4.
3. Keyting. op. cit. IV. 3.

PART VII

THE SECOND ALTERNATIVE FOR FULFILLING THE LIVING EXPERIENCE:
"I Realized God Through Mathematics"

CHAPTER 18

The Way:
The Analogical Field

A second alternative for following the living experience "I realized God through Mathematics," and which is based on another interesting aspect of the primordial thought of Dr. Raynaud de la Ferrière, is the analogical field which also guides us in the participation in his conclusions and experiences.

This field is so vast we are going to consider it in depth for following the living experience which is the theme of this book, considering that all the contributions of the analogical systems are defenders for it by themselves, better by their straightforward methodology and direction, or in their modern systems which escape experience. Every manner of analogical thought helps to bind the concordances, the associations of ideas from the origin to the universal. Dr. Serge Raynaud de la Ferrière uses them frequently throughout his 100 works, bestsellers in Latin America. And from there we can extend ourselves with diversity in the broad world of the analogical wisdom for letting us obtain a cosmovision of possibilities of concordance. This permits in a certain manner a direct way in this second alternative of following his living experience which we are mentioning.

It deals with the analogical thought united to the methods of concordances, the concordant variations, a little in the manner of John Stuart Mill who said that, "If two or more cases of phenomena investigated have a unique circumstance in common, analogy can exist only in the circumstance in which all the cases agree on the cause, or the effect, of the given phenomenon."

We find also a reflection of Norbert Wiener that places mathematics as a platform of a possible realization of that which Dr. Raynaud de la Ferrière establishes in his observations.

A deepening of the analogical thought, brings especially the theories to be placed around the mathematical models.

Thus within the principles of correspondence, Master Serge Raynaud de la Ferrière soars to different horizons of his vision in his analogical verifications, frequently around mathematical criteria:

> "Without a doubt, the science of numbers constitutes a study very beautiful. I seemed attracted to it from my youth, thanks to my studies of analogy between the macrocosm and the microcosm. In truth, everything in the Great World happens exactly like that in the small world. What happens in the universe also happens in the human being (universe in miniature).
>
> "The sun retrogrades in its apparent movement, one degree for each 72 years (equinoctial precession), and the same in the human body. There are 72 beats per minute, which indicates good health. The heart beats four times while respirating one time, that is to say, there are 18 respirations per minute, which corresponds to the 18 years of the nutation of the terrestrial axis under the lunar influence. (Remember that the moon turns through four phases.) The 25,920 respirations in 24 hours make you think about the apparent time which the sun needs in the equinoctial precession for returning to its initial position. The great year of Plato (365 x 25,920) is too important not to mention, albeit briefly."

We know that analogy, (from Greek analogos, proportion), is, in a mathematical sense, the equality of the relation which joins, two by two, the terms of distinct binas. Analogy is also the concordance or likeness between two or more terms in the function of their common relation with a third; life is like a river because both flow. Thus the scientific operations would be considered in good part as the transformation of analogies in concept. Yet it can be said that, more than the same experience, it is the transformation of the analogy into concept which produces the scientific understanding starting from the ideation.

Nevertheless, we know that analogy, used in the manner in which the disorder of the phenomena begin to be structured through the recognizing of the affinities, regains a primordial role in scientific discovery, for example, in cybernetics where it fits models to the study of man on everything in the theory of information.

Abraham A. Moles, author of the informational theory of perception, says that the birth of cybernetics occurred with the publication of the celebrated book by Norbert Wiener in 1948. He qualifies cybernetics as a crossroads science because it is the point of convergence of the theory of communications, of economics, of nervous physiology, of applied mathematics, of the theory of servomechanisms, of the science of government, which has given it its name. And we would add our theory of cosmobiology, intended as cosmic biocybernetics applied to 12 areas of basic specialties.

Abraham A. Moles says that cybernetics will be defined as a general science of organisms, independent from the physical nature of the organs of which they consist. It takes what is common to all organisms; it uses analogy as a method of investigation and as a principle of argument. Cybernetics consumes the break with anthropomorphic classifications of the sciences, be it in a function of the sensory channels (optic, acoustic, etc), or in a function of the origin of the organisms, (sciences of nature, of life, human sciences), in order to restore structural analogy as a new key of understanding of the world which surrounds us. The ordinary task of cybernetics is to define the level on which it is placed, whether it deals with a nervous system, in an automobile or in a community, an atom of structure which is called quadrupole, following the others, such as in a system of television. In this, the atom of structure is a case with an entrance, an exit, and a function of transformation. It does that for discerning the scale of magnitude of the 'cases' which it will need to group for constructing an intelligible model. Thus .it establishes a hierarchy of structures which then permit changing at will from a level without changing method; and thus discerning in each level considered, repertories of elements, categories of signs, or atoms of structures, as a homeostatic regulation. There are apparatuses which can be self governed, like the celebrated electronic tortoises of Grey Walter, which, constructed in 1950, would recharge their batteries upon weakening their tension, or can repair and even reproduce themselves in determined conditions as set forth by one of the major mathematicians at the end of the recently terminated era, Von Newmann.

Abraham A. Moles adds that the hierarchicazation of the levels causes the subordination to appear of a multiplicity of elements in a global structure. The other dialectic pole of the science of organisms is, then, the notion of a set or a totality, this being defined not as simple totalization, but as reality distinct from the sum of the parts which integrate it. It was necessary, then, to elaborate a theoretic concept that gave an account of this new dimension of the world of organisms; it is the notion of complexity, univeral measure of the technological universe.

There arises then the concept of complexity which is different from complication, since it is called complicated if it is a system of distinct organs, and complex when it is composed of a great number of similar organs and connected in an identical manner. A calculating machine is an instrument for overcoming complication, for through the complexity of its structure it permits reducing the complication of the initial whole. Thus the complexity has a creative power; since man, through the technological universe, becomes incapable of foreseeing all the potentialities of the set of elements and their analogies, as we shall see.

CHAPTER 19

The Reality of Analogies in the Global Structure—Archeometry

Again, the reality of analogies is established where the signs are articulated in a global structure of language within the whole, be it mechanical, biological or of complex social type, integrated in more or less broad structures, actuating each part after part and at the same time receiving the actions of these. The organism itself appears as a super-sign, bearer of a certain quantity of information, in an hierarchical process which constitutes one of the properties of the intelligible forms of the Gestalt system. Reality overwhelms us by its richness of analogies and seems to exceed the possibilities of our spirit. It is cybernetics which has restored the structural or functional analogy as the key to the world around us.

The importance granted to analogy orients the thought of technical phenomena towards a philosophy of *as if*. That which is 'the world in itself' escapes us; and perhaps the expression 'the world in itself' lacks sense. But new technical thought makes 'as if' the world had direction, as if its analogical constructions were true, and this they are, since they triumph.

It is true that Ernst Mach defines analogy as the relation between systems of homologous concepts which can give rise to difference or coincidence.

On the other hand, in the measure in which it has returned to dominate the ontological concern, analogy has been used lately in the sense of first philosophy: the meeting with the Supreme Cause. But let's look at mathematical ontology.

Let's say, for our part, that on the epistemological basis of a mathematical ontology, we are three-dimensional beings who interpret 'attraction' as a force of gravity, after the point of view of Einstein: 'We are beings who, if we lived in a fourth dimension, we could say that we would face an alteration in spatial geometry of four dimensions.'

In the preface of the *Third Handbook of the Scientific Theory of Cosmobiology*, we have pointed out the thought of David Halliday and Robert Resnick of the University of Pittsburgh and the Reneuber Polytechnic Institute, respectively, who say:

> "We ourselves are a mass of nucleii and electrons bound together in a stable configuration through Coulombian forces. The importance of the Law of Coulomb goes much farther than the description of the forces which extend between spheres and charged rods. On incorporating this law into the structure of quantum physics it describes correctly, (a) the electrical forces which unite the electrons of an atom to its nucleus, (b) the forces which unite the atoms between themselves to form molecules, and (c) the forces which bind the atoms to the molecules among themselves to form solids or liquids. Thus the majority of the forces of our daily experience are not gravitational but electric."

We could add the principle of correspondence of Niels Bohr, in accordance with which the radiation emitted by a quantified system should coincide in frequency, intensity and polarization with the correspondence to a classic oscillator for quantum numbers sufficiently large enough.

And as if to corroborate once more the fulfillment of a truth through mathematics, we have, with respect to this prospective, the formulation of Gaston Berger, (1896-1960) engineer and Director of Higher Education in 1952, who dedicated the last years of his life to setting the bases of a science of action.

Berger defines the prospective as that attitude which consists of 'forming the present starting from the future, instead of considering it as a secretion of the past.' The prospective tends to rule our actions by the future, to make the acts depend on their consequences to increase the correlation between the present of our perceptions and the future of our projects, in a world which has widened its field of vision of the future through statistics and the machines appropriate for manipulating complexity. The appearance of this concept shows clearly that we live a new rationality which tries to form unheard of systems of thought, such as cosmobiology, though the help which the institutions contribute to us, (history and experience) appears effective if it informs it from current technical rationalism, which is found in applied mathematics and in formal logic.

The sciences themselves have experienced a change in both signification and value. Mathematics now performs an important role in technological life, especially since theoretic thought seems supported by the possibility of giving to such thought a numeric form and of providing the result which the ordainer brings about. The mathematician thus appears free from the inferiority complex which he felt concerning both the physicist and the engineer, inasmuch as he also becomes a performer.

The powers of imagination are common to scientific and artistic thought which now occupy the primary position in the scenario of our mental activities.

Hence it follows that our ideological frameworks, our perspectives, our material inclinations, and the same 'field of possibilities' will change very quickly. What was true yesterday is no longer true today, and will be much less so tomorrow. We are assisting in a 'flight towards the future' of our civilization, which puts off difficulties until tomorrow in the hope that they will make up their minds. Any method of action does not pass before being already a speculation about the future. We find ourselves, then, in a new situation, where what is going to not be is already the normal and necessary product of that which is now. Where tomorrow's solution is not the solution of today; and where the active man already cannot be satisfied with tradition, nor with the information received from the present and from his own experience, but from a new age.

From this fact, the prospective does not exhaust the future. On the contrary, it possesses a serious function. It is an antifatalism which is unable, nor does it want, to substitute the predicament from the past for an oppression of the future; and the determination of acts by their consequences at the time of their antecedents does not hinder in any way the richness of what will be. This is a transcendent fact.

When each man receives an education which permits him to handle the problems that he must study (make up his schedules, make decisions), he takes on an immense task of integration in our culture, that is to say, in our language, of activity and of technical thought, profoundly ignored and even slighted by traditional humanism.

For this reason, the technological man turns his thought to the biologists and geneticists to obtain the spiritual supplement (or capacity for integration) that will permit them to remove the opposition between profundity and what is superficiality which constitutes the tragedy of our time, as Abraham Moles concludes. And we consider, for our part, that this will be resolved in good measure through cosmobiology as the cosmocybernetics which, with its mathematical base applied to its applied sciences and sciences of contribution, provides the prospective which Dr. Raynaud de la Ferrière designates as a 'science of sciences.' Thus, Dr. Raynaud de la Ferrière says about the analogies that 'mathematics, geometry, physics, and astronomy fulfill an important role for those who wish to reform the world. There are bases, solid and concrete, to be established in the material world before elaborating a better spiritual universe. . . ."[1]

Let us pause now to bring our attention to the ancient thought which Dr. Raynaud de la Ferrière views from different angles and which it is convenient to refer to at this point. This will permit us to extend the analogical way which allows us, through mathematics, to unite concordances toward the Supreme Cause, since the archeometry which Dr. Raynaud de la Ferrière mentions implies kinetic uses in accordance with the mathematical models of the modern cosmic biocybernetics previously mentioned. He said:

> "Archeometry says, ontologically, that life produces the organism which manifests it, and not the contrary. That it is the organizer according to the specific conditions of the beings and the material means."[2]

Farther on, Dr. Raynaud de la Ferrière makes remembrance of archeometry:

"Everything always rests on good solid bases, it has to do with making the search of analogies, of similitudes, and of symbols, for laws capable of functioning on diverse planes; briefly, action and reaction, cause and effect, and above all, a mechanism that is not always 'visible' at first sight but which reason proves to be well founded."[3]

"Archeometry is a material translation of the word in form, color, fragrance, resonance, and flavor. It is a synthetic figure in circular form, divided into concentric zones which contain the correspondence of names, letters, notes, the zodiac, the planets, the alphabets, (Vattam, Hebrew, Latin, Samaritan, Chaldean, Assyrian), vibrations in correspondence, such as a metric pattern which reforms sonometry and which can serve as a rule for aesthetic proportions of all things traced graphically.

"The primary colors, yellow, red, and blue, are distributed in a triangle within the circle, in such a way that white is always found, theoretically, existing at 180 degrees to each of these three colors. The result, then, of this is 360 definite tones and each one of them is identified with an ordered number which permits rapid discovery of its exact composition. Certainly the relations of music are written there according to the physical method, inversely proportional to the pattern of current measurement. (D-flat = 100,000 or 1 meter; the figure 100,000 = 625 x 160) The outline of a form, whatever it may be, comes to be thus an expression of the key of a universal canon. That is to say, that in accordance with one name or one idea (or a sound or color) it becomes possible to obtain a design, a graph, an entire structure with the exact proportions as regards width, height, etc. that has to do with an aesthetic monument, a religious building, etc. . . ."

"By an inverse procedure, they are likewise adapted to literature, colors, forms, etc."[4]

"Thus it becomes easy to notice the alterations which a symbol has undergone, since archeometry represents a special spherical plan, which manifests to us a scientific diagram of the cosmological and morphological correspondences based on the word and its equivalents.

"The harmonic laws of proportions, of forms, and of music are the same. Thus, a sacred object, a church column, a lodge of initiation, a temple in its totality, all require a metrical synthesis which evokes the name towards which the aspiration ascends, the symbol, the offering, the ideal, etc. . . ."[5]

" 'Arka' is a word understood as 'Revelation' in the mysteries of the 'Son,' as He is the Word, and Creator through the word. He is the same word manifested by numbers and rhythms. Further, He is the poem of the Word. 'Matra' is the supreme measure of the Principle. Measure-Mother is what lives in the Word-God, as all His creative thoughts. It is she who manifests the unity in all things through the universality of her internal proportions." "Archeometry (Arka-Matra) properly, then is the true Principle, located in the rational method for establishing a system of manifestation of sacred thought."

"Archeometry is an instrument from which the ancients established all the esoteric myths of religions. It is the canon of ancient art in its diverse architectonic manifestations, be they musicals, poems, or theogonies. It is the sky which speaks: each star, each constellation becomes a letter or a phrase or a divine name illuminated with new irradiation, the ancient traditions of all peoples.

". . . it is the key of sonometry and provides colors for the painter and, at the same time, forms for the architect. It does not try verifying with pre-established instructions, but investigates the manner of making expressions arise in accordance with a definite method; it builds nothing but it presents the principles which it desires to comply with for beginning these or those outlines.

"It restores the arts to their synthesis, re-adapting expression to the true sources. It provides the key to successive religious and scientific adaptations of antiquity. It allows the architect to construct forms in accordance with a name, an idea, a certain color, respecting the vibrations emanated from a harmonic curve, since they correspond to a planetary influence related to the origin of revelation."[7]

And in another text he adds, concerning the thirteenth century:

"He speaks in that time (XIII Century) of a rather strange personage, and intrepid traveller, and no less indefatigable writer, well-versed in the Arab philosophers who spread Averroism everywhere. For combating it, the Franciscan tertiary inflamed with love for Christ like Francis of Assissi, Raymond Lull, a native of the Balearic Islands, theologian, apostle, missionary and mystic, his teaching would have repercussions on various philosophical schools."

Certainly we were able to consult about his last work, *Theatre of the World*, in its original edition of 1606 in the library of the Convent of Ocopa. With regard to Raymond Lull, we have also referred to his work *Ars Magna* as well as studies of the eminent Lullists, the Frenchman Raymond d'Alos-Moner, and the Spaniard Joaquín Xirau. As concerns Juan Pablo Gallucio, it is he who presents a cosmological instrument with superimposed discs carrying the movement of the multiple facts about celestial concordances.

In the study of Raymond d'Alos-Moner, he points out, "It is possible to say that the true Raymond Lull has been unknown until our own time."

In *Ars Magna*, Raymond Lull says:

"It is necessary to detach primitive ideas—the 'dignities'—principle of all possible combination and to discover the laws of their possible and necessary combinations. In possession of both these things—the ideas and laws of its possible combination—we will have, in principle, the key to the world and of the necessary relations of the world with God. By means of an adequate technique, we will find ourselves in conditions of realizing all the possible combinations and of deducing in the intersection of them, all the possible worlds and, among them, the true one, since the true world is, evidently, one of the possible worlds. It is the idea of the combination. In its intimate architecture, we will find the longed-for 'ladder' which permits us to ascend by necessary reason to the source of all reason, and to descend from there to its infinite branches."[8]

Afterwards, Joaquín Xirau comments:

"There will be no other, in the whole 17th century, but the Leibnitzian idea of the analytic combinatorial machine. And from the universal characteristic which they were obliged to serve, according to Leibnitz, as algebraic instrument for the reduction of all equations, he declares in an explicit and repeated manner his Lullian source. Such is the origin of contemporary algorithmic logic.

"Assembled in an harmonious equilibrium around the axis of Franciscan cosmology, his interest was found in a resolved eagerness for the unitary and hierarchical classification and of which its totality is considered better than a definite conclusion, as a point of departure for the work of knowledge.

"Based on this, it is necessary to 'investigate,' show, and state all the secrets of nature.

"With the help of art and the sixteen trees, all sciences can be studied, but, only in summary, in an abbreviated and potential form. Its explicit development remains reserved to the future, and is found, in principle, infallibly guaranteed by the rigorous combinatorial inventiveness implied in the methodical unity of art."[9]

Let us see, then, that resorting to the direction of the ancients in the analogical thought in union with the moderns leads to a globalizing methodology that well explains this second alternative in that mathematics is seen accompanied by the universal concordances, which bring us to the theme of the following chapter, referring to analogy and the universe.

NOTES

1. The Black Book of Freemasonry. 102/7/1.
2. Psychological Talks XIX. 18/3.
3. ibid. XIX. 31 ult.
4. Black Book of Freemasonry. 103/2 to 103/4/2.
5. ibid. 101/ant. pen to 102/1.
6. Psychological Talks. XIX. 16/2.
7. ibid. XIX. 16/2.
8. Raymundus Lullus, Ars Magna. 90/2.
9. Joaquín Xirau, Life and Work of Raymond Lull. 41/1.

CHAPTER 20

Analogy and the Universe

The sense of analogic thought achieves extension and depth in the discovery of its generic concordances, which introduce themselves among different values. They can start from logic and science and arrive in profundity, or extension, at the brink contemplated by contemporary philosophy, as in the analogies from attribution of proportionality; or in the style of Locke who includes analogy among the degrees of assent and considers it precisely as the probability which concerns the things which transcend experience.

Analogy, according to Locke, is the only help which we have to achieve probable knowledge of the 'finite immaterial beings outside ourselves,' or of 'material beings which, due to their distance from us, our senses are not capable of perceiving,' or, finally, 'that which refers to the *modus operandi* in the greater part of the works of nature,' which is hidden from direct human experience.[1]

Leibnitz agreed with Locke in seeing analogy as 'the great rule of probability,' as soon as that which cannot be witnessed by experience can be considered probable if it is found more or less in agreement with the established truth. Leibnitz adds an example regarding the use which men of science have made of analogy and recalls that Huygens, basing his beliefs precisely on this, judged that the state of the other planets is similar to that of earth except for the differences produced by their relative distances from the sun.

In fact, men of science during the 17th and 18th centuries made great use of analogy, and Kant has used the term, not without good reason, to express some fundamental principles which regulated the science of his time.

Kant understood by analogy a form of theoretic proof, and defined it as 'the identity of the relation between principle and consequences, (between causes and effects), inasmuch as it takes place, notwithstanding the specific differences of the things or of the qualities in themselves (that is to say, considered apart from the said relation), which contain the principles of similar consequences.' Kant enumerates three 'analogies of experience,' which he indicates in the following manner:

> "a) The principle of permanence of substance, which is expressed by saying, 'In each change of the phenomena, the substance remains, and the quality thereof in nature neither increases nor decreases.' b) The principle of the temporal series, according to the law of causality which is expressed thus: 'All changes occur according to the law of connection between cause and effect.' c) The principle of simultaneity according to the law of reciprocal action which is expressed saying: 'All substances, inasmuch as they can be perceived as simultaneous in space, find themselves in universal reciprocal action."

Kant has clarified the sense according to which these principles are called analogous, in mathematics. Analogies are formulas which express the equality of two quantitative relations and always produce constituents, that is to say, when three members are given in proportion, the fourth is given also, which therefore can be constituted.

In philosophy, on the other hand, analogy is the equality between two relations, not quantitative, but qualitative relations which means that if three terms of the proportion are given, the fourth term is not given with them, but the only given thing is a certain relation between them. This relation constitutes a rule to search for it in experience, and a sign by which it can be discovered. In this way, the principle of permanence of substance, the principle of causality, and the principle of reciprocal action do not truly constitute the objects of experience, but are useful only to discover them, and place them in the universal order of nature. These principles are a priori and, therefore, indubitably certain, but at the same time they are devoid of intuitive evidence, since the 'axioms of intuition and the anticipations of perception are constitutive principles because they teach how phenomena, be it in that which refers to its intuition, or be it with reference to its perceived reality, can be produced according to the rules of a mathematical synthesis.'[2]

One of the analogical procedures consists in the creation of symbols which have a greater or lesser likeness with real situations, and whose relations produce the inherents to the elements of such situations. Such symbols result at times in mechanical models, or patterns, schema, or machines which reproduce the relations in which the true elements are in the middle, such as they are; for example, the models of the solar system, of the structure of the atom, of the nervous system, etc. In other instances, such models are obtained through the process known as extrapolation which consists of taking to the limit the development of a set of cases ordered in a series, in which it is supposed that troublesome influences have been eliminated. Thus, for example, we speak of infinite velocity or zero velocity, of masses reduced to a geometric point, of perfect levers, of ideal gasses, and so forth. Each model constitutes an example of analogy in the first sense, since the characteristic of a model is to reproduce, among the elements themselves, the same relations of the elements as in the true situation. But physicists speak at present also of analogy as of conditions or integrating elements of hypothesis, and of scientific theories. According to this direction, analogy enters into the constitution of a hypothesis and ought to be analogous to some known laws, in this sense, analogy corresponds in physics to that which is the musical sense in music; it will guarantee the appropriateness of a scientific hypothesis to the uniformities expressed or formulated by the laws.

A phenomenon which varies in some ways each time that another phenomenon also varies in some particular way, is the cause or effect of this phenomenon or is related to it through some fact of causation.

Mach reduces all scientific procedure to this method, namely 'the method of variations, which consists in studying, in each element, the variation allied to the variation of each one of the other elements. It matters little that those variations are produced by themselves or are provoked voluntarily; the relations will be found through observation or through experiment.'

Bohr and Heisenberg have shown that different descriptions are possible for the same phenomenon inasmuch as the limit between the observing system and the observed system is not rigid, since all observation, be it natural or experimental, presents the division between the observing system and the observed system. The validity of this division has been proven and confirmed by quantum physics, with reference to the relations of indetermination, that is to say, the action which the observing system exercises upon the one that is observed.

Heisenberg comments: "It would be possible to treat quantitatively the chain of causes and effects only when the totality of the universe is considered as the observed system; but then physics will disappear and only a mathematical framework would be left."

According to the thought of Heisenberg, upon the totality of the universe being observed as an observed system, physics would disappear, and we would include the disappearance of concordances and symbols, leaving us in the purest mathematics, before the same direct way for the universal interweaving of mathematical thought with causal thought: causality and mathematics.

Analogies and calculation, concordance and mathematical models, cause and number, are produced through mathematics. Let us now look at another of the ways proposed for achieving this objective: the epistemology of life.

NOTES

1. John Locke, Essays. IV, 16/12.
2. Immanuel Kant, Critique of Pure Reason. Anal. princ. III, 3.

PART VIII

THE THIRD ALTERNATIVE FOR FULFILLING THE LIVING EXPERIENCE:
"I Realized God Through Mathematics"

CHAPTER 21

The Way:
The Epistemology of Life

I. Truly, from the mathematical observation of universal reality, the living experience arises: 'I realized God through Mathematics.' It was a consequence which would have to arise to parallel therefore, a concept of divinity which Dr. Serge Raynaud de la Ferrière himself wishes to establish: a logical, archeometrical, epistemological concept which will imply the embracing of diverse religious concepts from the primitive cults to the more complex religions, with an explanation appropriate to the era of true investigations.

When Walter R. Fuchs expresses that mathematics as an activity is a natural thought in which the mathematician makes of truth a direct vision of the dynamism in which all being consists, he confirms the horizons and discoveries of the ontology which, for his part, Dr. Raynaud de la Ferrière undertakes towards the farther shores in which being carries its abstract contents of experience. That is to say, where logic is extended towards the ultimate causes, entering into the investigations of philosophy, and of the mystery of Being, without abandoning the causal and scientific platform, thus presenting one of the more revolutionary demonstrations of the thought of a new Age. It includes in its incursion the thorough analysis of a new concept of the Divinity which 'removes the veil of superstition from the decadent religions to reveal the profundity of cosmic knowledge which unifies the true philosophical principles in the unity of eternal truth.'

We accompany him, then, in such elevation as in a text for meditation which is pronounced in a Central Temple of universal dimensions:

> The conception of the Divinity which we want to establish is:
>
> > the Supreme Cause which consists of two divisions: the non-manifested effect and the manifestation;[1]
> >
> > the Supreme Cause which in its infinite essence, naturally escapes analysis, but different manifestations themselves can be analyzed, with the object of leading the spirit to a better understanding of sacred things, for which reason it is necessary to focus on a base of an epistemology of life.[2]

II. It is important to stop at this point and outline the Third Way. When Dr. Raynaud de la Ferrière says that it is necessary to have an epistemology of life to analyze the manifestations of a Supreme Cause, he implies the fundamental incursion of a discipline which has as its object the study of the forming and transformation of scientific concepts. This, within the definition of a new concept of divinity, and as a basis for approaching a better understanding of sacred things, reveals the illumination of a new way of thinking for a new humanity. It is a new way of feeling for a new subjective realization of universal existence with a basis in scientific thought, a mysticism based on science.

Through its transcendence, let us pause long enough to determine what the French investigator, Francois Guéry says: 'Although the word "epistemology" is used in English very often in the generic sense of theory of knowledge, of gnoseology, it is understood in the wider philosophic sense and not only in the scientific sense. In the modern French sense, by 'epistemology' is understood the discipline which has as its object the study of the formation and transformation of scientific concepts; how they interchange from one science to another, how the field of a science is constituted; according to which rules the said concepts are reorganized through successive changes, and how, when referred to its own rules, a scientific practice becomes conscious of its method.'

Epistemology—before philosophy, as Jean Piaget says, goes back to Plato—has been placed at the side of the sciences for guaranteeing their validity.

For the Bachelardian current in contemporary epistemology, the sciences are those which are 'from the side of philosophy,' and which present the same dialectical properties which reason has for the philosophers.

Gastón Bachelard outlines a special affinity with Dr. Serge Raynaud de la Ferrière in the value of the syntheses, the mathematical recreations, and the sudden animation which the syntheses of physics and mathematics give to the soul.

And he testifies to one of the more typical experiences of wisdom upon explaining: '. . . becoming totally absorbed in scientific research with all the forces of life.'

And he adds, in *A New Scientific Spirit:*

> "By what light is the value of the sudden syntheses first recognized? Through an unspeakable clarity which gives our reason security and happiness. This intellectual happiness is the first sign of progress. It is the case recorded here, with the phenomenologist Jean Hering, (*Phénoménologie et Philosophie religieuses*. Strasbourg, 1925, p. 126): ". . . that the person who has evolved the most, due to the greater extent of his horizon, will always be in a condition to understand those who are less . . . while the contrary is not possible." Understanding has a dynamic axis which is a spiritual impulse, a vital impulse. Einsteinian mechanics add something to the understanding of Newtonian concepts. Current physics determines a synthesis between the last two groups of concepts which are developing and completing the Cartesian epistemology. If it were known how to duplicate the objective culture by psychological culture, completely absorbing in the scientific research all the forces of life, one would feel the sudden animation which the creative syntheses of physics and mathematics gives to the soul.

Is it perhaps due to this receptivity in the face of the novel that Bachelard could be christened by J. Hyppolite, in an article in the *Review of the Sciences*, as the authentic founder of contemporary epistemology? asks Francois Guéry. And he answers: "To judge the character of novelty and difference of this scientific discipline, it is not sufficient to refer to the simultaneous and correlative progress of various branches of the natural sciences and of mathematics. It is required to take particular care with the principal orientations of the transformation of the philosophy of science which, being in difficulties due to the fact of its unforeseen displeasures, must be reformed—according to Bachelard's own expression—from its very principles."

In effect, he adds, this novelty which makes epistemology an autonomous scientific discipline is marked out on the basis of a philosophy even to an unchanging metaphysical tradition dating from classical times. The great themes of Kantian philosophy are found in what underlies the epistemology of Meyerson at the beginning of this century, and in which he thought to break with all the points of Kantianism, as is Bergsonism; while starting from Bachelard, the perspective changes fundamentally.

Though in the new era of true investigations, as Dr. Raynaud de la Ferrière calls it, the essential scientific change has taken place in the natural sciences, one can examine this revolution, as Guéry says, in geometry, for example. This is so, not only through itself, but through the conception of physical space formerly alien to it; in the realms of chemistry, physics, astronomy, and optics, realms re-elaborated by modern atomism and where relativistic mechanisms were born, quantum mechanics and wave mechanics.

Also affected in this new era of wisdom is all of physical realism, just as they leave the relations between material reality and abstract thought modified, in the hope of explicitly drawing from there the appropriate philosophical lesson. Experience and formalization are going to be placed in a fundamentally new light, one with regard to the other. As regards this point, the simple fact that the theory of relativity does not arise from direct reflection on the objectivity of the world, is somewhat characteristic.

Bachelard insists, throughout his expositions relative to the 'new scientific spirit,' on the abstract aspect of the decisive progresses of contemporary sciences. In fact, the use of the basic principles has as its result, as in the case of geometry, the invalidating of an intuition which was given as evidence. This step is not made except with the total disappearance of the reference of elaborate science with the immediate reality from which it is freed. From this arises the new status of scientific reality.

By a dialectic inversion of perspective, adds Guéry, the world is no longer that which verifies science. On the contrary, it is the image in perpetual change which science offers to us at the moment of its verification. It is 'our verification,' that is to say, that of the researcher and of the scientist, as Bachelard says. And thus, the Quantum Theory, the pure gratuitous initiative of the mathematicians—as a realist would say—represents a physical elaboration so far removed from common sense that it requires a genuine education, an education of the spirit so that one succeeds in representing it clearly in the world of common experience.

We could consider that the elaborations of Einstein, Planck, Maxwell, Oppenheimer, Poincaré, and other great renewers of scientific thought, and therefore the contemporary ideation of truth, were the forerunners of this aspect of the thought of Serge Raynaud de la Ferrière.

Certainly, instructed by the Einsteinian relativism, Bachelard shows without difficulty that science has known how tô pass the limits which, far from being inherent in them, only constitute a provisional state of their development. Following Brunschvicg, he affirms that time and space appear organically placed in a new mechanism, and that the idea of a time of ideal reference, or identical to itself, is in need of sense. The relativism has to be new to integrate in a system of the known object, the very place of the physical observer with respect to the object.

Prior to the appearance of the topocentric mathematical models of the discoverer of topocentricity, the Hungarian mathematician Wendel Polich, this Bachelardian situating of place itself on the plane of the observer is parallel to the cosmobiological vision of Dr. Raynaud de la Ferrière when he says in *The Great Messages*,

> "On the contrary, the sensitive, essential organ of the microcosm is found divided into four arcs of 90 degrees, and each one of these quadrants is divided into three equal zones of 30 degrees, which are the Cosmobio-

> logical Houses; that is to say, the specialized realms of our receptivity to the influences of the heavens. If one speaks always of the inequality of the Houses, we know that it is only by the construction of the imaginary planes, thanks to the projection on the equator (which is not its base). The ancients conceived the Cosmograms traced on the upright, and all the aridity of this question is abolished with the stereographic projection of our sphere."[3]

In relation to the ascensional calculation of time for the given place, a science of time appears, a science of the inter-variable cycles, Cosmobiology. We see that it is now possible, (thanks to the mathematical models which computer science permits). The pressure on the topocentric and biocentric place the variability of the cycles in the phenomena of biology and in human events; just as finding constants permits uses and diagnoses in diverse sciences, from sociology and the lot of biomedical sciences, to the historical investigation in which is applied in remote parameters, even the geochronology of the Pleistocene Age. This is done by the geologist Frederick Zeuner, as presented in the *Scientific Theory of Cosmobiology*, Acts of the Sciences I.

All this demonstrates not only a new scientific spirit, in the manner of Bachelard, but the dawning of a new era of controls and unsuspected plans.

The same occurs, by the way, in regard to the identification of the effect with the cause. Instead of reducing the effect to the cause, the process of contemporary science consists in enriching the cause with the effect, and in showing that the new complexity introduced by causality is irreversible and that it contains the simple, initial thing as the poor abstraction drawn out of a rich whole. So then, the process of science does not consist in the fusion of two things into one (from effect and cause) but in the division of one thing into two, according to a certain evidently dialectic schema, in specialization or specification of the concept, and in the division of science itself into ever new areas.

So then, when Serge Raynaud de la Ferrière considers the concept of the Supreme Cause containing two divisions, he observes, contemporaneously to Bachelard, an evidently dialectic schema, which leads through synecdoche to a transformation of mysticism.

NOTES

1. Series of Great Messages. 285.
2. ibid. 414.
3. ibid. p. 77.

CHAPTER 22

The Transformation of Mysticism

It coincides with the clarity and interest in this new concept of divinity of Dr. Raynaud de la Ferrière which permits a sense of causality to be applied to the supreme notions of an intelligent principle, proper to the new age, and which permits the transformation of mysticism in its conceptual and doctrinal bases.

This is a little like what he sets forth:

"The Supreme Cause includes two divisions: the non-manifested effect and the manifestation. This expresses the dialectic conclusion of a causality which is diversified in a passive effect and an active effect, which he also applies indicating that we easily conceive that man, engrossed in a dream, or in full activity, expresses himself in two phases of the same individuality. The same occurs in all the ethers with regard to the distinct manifestations of the same entity.

"The two phases of the Great All are identical in essence and only have differences of state and nature. These are two phases: the static principle called ABSOLUTE, and the active principle, called GOD.

"The one is in rest; the other in movement: The God Non-Manifested and the God Manifested . . .

"Now one can better understand that the Trinity of the Manifested One is also the reverse of its own dynamic activity characterized by three movements. This is the Divine Trinity, which is found in the basis of all the great religions . . . The first phase of the divine CONSCIOUSNESS-IN-ITSELF is Breath, Life, Light (God the Father).

"On a physical plane one can cite this correspondence in the awakening of the conscience which operates before the bodily faculties awaken.

"Immediately comes wisdom, and love, and harmony; it is the Son. In this one are the different doctrines, that is, in which some see this unawakened attribute, Space-Time, the matrix of a future Cosmos, that is, the Celestial Mother, or that which Genesis expresses in 'darkness reigned over the face of the abyss.' Other religions see the attribute already awake, (the Waters, above which the Divine Spirit is moved), as Genesis mentions it, regulated by the Law of Action, that is, the Son of the Divine Thought.

"At any rate, Son or Virgin (Mother), there is no difference of state, being always the second manifestation.

"Then comes Thought in Action, the third phase of the awakening of the Active Principle. It is the Ideation: the Holy Spirit.

"In short, Father, Son, and Holy Ghost, or, if one prefers, Energy, Form (which is the matter in its manifestation), and Thought (polarized force) can be defined as Spirit, Love, Duty, or Will, Wisdom, Activity, according to the philosophy or doctrine being professed . . . but in any case, it is seen that, in spite of the apparent difference, they are nothing but differences in terminology and that, in essence, it would be simpler to understand the value of the words, in the study of the great causes (currents of thought) that inspire the world."

On the other hand, the power of science must allow the fusion of the sciences which occupies, in the thought of Dr. S. R. de la Ferrière, a primordial plane as one of the bases of the union and greater understanding of men, a world united by knowledge overflowing in the impregnation of an axiology of values of approach and universal understanding, which must favor the transformation of mysticism in the union of religions.

When Master Raynaud de la Ferrière says that the Supreme Cause contains the non-manifested effect and the manifestation and Life-Form-Thought, an evident transcendence of this concept of divinity founded in a causality, in a dialectic of causality, the application of an all-encompassing concept is made to all the religious principles, both primitive and civilized, with a unifying power.

That is to say, an epistemology which associates at the level of perception, (the sensible, the appearances, the common experience), and the level of understanding (science, the concepts, the laws, deduction) is a synthesis which includes a metaphysical activity of dialectic reason.

Thus the progresses of the spirit are better, as is the knowledge of the world which finds its definition and its own notion through the development of scientific thought.

What really is the spiritual realization? Among other things, as Bachelard says, "the sum of proven ideas"; it is the verification of the scientist, the true. It is necessary to see in it an authentic Bachelardian definition: 'never is it what one was able to believe'; it is 'always that which one ought to have thought'; belonging to an era of knowledge and not from simple believing like the era that has passed.

Master Raynaud de la Ferrière says, for his part, in *A Mathesis of Psychology:*

> "Life is a chain of experiences, and the synthesis of its different actions constitutes spiritual fulfillment.
>
> "Intellectual knowledge is indispensible for having an exact idea of the universal mechanism with which we will be able to elevate our spirit through pure inspiration. This objective and positive root of science, or at least of knowledge, is the basis of the foundation of the most subjective edifice which we wish to construct for arriving at God.
>
> "That is to say: is science necessary for fulfillment? Is there in wisdom that which is the Truth? It is not a question of answering yes or no, as long as it is not possible to conceive reality outside of the All, and one is not able to admit the absence of knowledge except outside of the Light. For fulfillment, one must know of the All, and the Truth is within the Light.

"This reiterates that there is need of a synthesis of knowledge and of wisdom. Many who have been educated are not wise and many who are wise have not been educated.

"Truth is the Eternal Word. The wise man of times past perhaps was not like the wise man of our time; the educated man of the Middle Ages not like the 20th century scientist . . . but Truth is eternal and as a result is beyond the limits of learning and of wisdom: it is the synthesis of the two.

"Science and religion are the two indispensible polarities for arriving at the Truth. Through Science one can understand the unlimited sense of wisdom. It is knowledge of science called 'official' and knowledge of the other, called 'occult.' Through religion (from the Latin religare, to unite), one should not understand an organized church, but the true sense of the general reunion of the concepts of wisdom and of all the doctrines studied in a common point from the uniting of the Great Ancient Principles."

These concepts of high unity of Dr. Raynaud de la Ferrière take us at once from the internal and hidden part of science to the consciousness-of-self of science.

That is, from mathematics we depart, and to mathematics we return, after passing through the analogical realization of a Supreme Causality.

PART IX

REINTEGRATION

CHAPTER 23

From Mathematics We Leave, and to Mathematics We Return

Up to here we have given processes and frameworks of the living experience (*vivencia*), from the bond of mathematical-biological-physical structures to the framework of socio-collective aspirations which have been able to substantiate some approaches of the living experience of the Master, Serge Raynaud de la Ferrière, suitable for a man anywhere.

We are coming to the end of our subject. In no way have we attempted to include all the primordial thought of Dr. Serge Raynaud de la Ferrière through the experience, "I Realized God Through Mathematics."

We have not tried to include all the constituents of this experience. The subject itself is full of tributaries and derivatives which are not encompassible within the framework of this book.

We have made an incursion into the primordial thought of the Master for establishing ways and models of identification with him, for which mankind and the thinkers and students are able to examine panoramic differences of the process for coming to said experience. We have given those processes and the framework which surrounds them for contributing to the expansion of the popular conscience for union with science, in the plane of conceptions of the Supreme Cause and for the union of the compared religions. We have done this with a basis in scientific ideations which emanate from mathematical thought, and its ancient and contemporary resources. That is to say, some aspects of the transcendence of mathematical thought in universal thought.

Nevertheless, there is a need to mention some culminating points, and for that we begin this exegetic apex in this selection from the same text of the Master:

"We are going to end now in the same manner with the explanation of the supreme objective.

"However, as this objective is the result of all the searches, and of all the perfections, we will change our procedure a little, that is, adding more concrete explanations. For this, we are going to refer, before concluding, to the characteristics which the celebrated French entomologist, J. H. Fabré, indicates: 'While I study more, and while I observe more, the ruling Intelligence of the Universe appears more irradiant behind the mysteries of things, Geometry, that is, harmony of nature, presides over all; the placement of the scales on the pine, the spiral of the shell of a snail, the symmetry of a spider's web, the orbit of a planet. Thus it presents everywhere such wisdom, in the world of atoms, as in the world of the immensities.'

"Therefore, the scientific world is obliged to abandon the simplistic theory of transformism and to admit a superior dynamism with the acceptance of an Intelligent Principle.

"In the chapter on Intelligent Finalism of Evolution, E. Arnaud says, "The Law of the Cosmos is that of the idea which unites all the phenomena among themselves; a mathematical law, in truth, but who will know the Divine mathematics? The evolution through the ages of the mineral, vegetable, animal, and human species is the only one which reveals a thought directed toward an objective, and which constantly models the matter with a creative breath.

"But let us dispense with science to say clearly that it is the fear of losing the personality which has created religions, and the idea of saving the individuality which has given a reason for being to the other doctrines . . .

"The truth is that it is necessary to return; all the rest are no more than conceptions, hypothesis, and beliefs."

Nevertheless, he continues his dissertation including the Christian and Hindu theological language, without abandoning in any paragraph the scientific comparativity:

> "Truth with all its complicated assimilation of principles is, nevertheless, very simple in its mechanism. It follows a natural curve, a cyclic Law, an inspiration, a retention, and an expiration, a movement which we see everywhere.
>
> "Everything emanates from a point and therefore, it is very normal that everything returns to it.
>
> "The absolute (God Non-Manifested) is presented in LIFE (God Manifested—the Father); in FORM, (the second attribute, the Son); and in THOUGHT (the third manifestation, the Holy Spirit).[1]
>
> —Primarily God does not exist more than in potency.
>
> —In the ineffable Unity is the first person of the Trinity, or God-the-Father.
>
> —Afterwards, revealed to Himself, an intelligible world is created; He compares the thought to the universal reason; it is the second person of the Trinity, or God-the-Son;
>
> —Finally, He acts and produces. His will is exercised, and His thought is fulfilled outside of Himself: it is the third person, or the Holy-Spirit.[2]
>
> "After the necessary evolutions, the journey is finished through comprehension (concretization of the thought-state of the Holy Spirit), through realization (Christalization of the LOVE-Christic plane of the Son), through the reintegration (dematerialization of the Divine Laws-State of the Father for merging finally in the Supreme Cause).
>
> "It is necessary to climb the ladder in the same manner that the Principle came down by it, by means of the Trilogism: Brahma Creator (the first Creator, the Father, God Manifested); Vishnú the Sustainer (the Son, or Mother, the Virgin, Matter, Matrix); and Shiva the Destructor (the developer, the Holy Spirit, the Universal Soul).

> "The Involution-Evolution movement is not a regular balance only; there are many more oscillations with effluvia which designate the different states of evolution, and it is by them that the reintegration does not come automatically."[3]

Thus we say that dividing the Supreme Cause, the three manifestations, Life, Form, and Thought, are approached by realizing them, while from the point of view of the man who wishes to merge, that is, to climb up the ladder in the opposite way they are approached by making them sublime. Thus a being who has come to the reintegration has transcended Life (God Manifested, the Father) but remains in this state as an intermediary between God Manifested and Non-Manifested, and corresponds to the thought anterior to the first manifested projection of the absolute.

Returning to the Master, he continues, saying:

> "Always, according to the Law of Involution—Evolution, that which has left God ought to reintegrate in Him. The principle is going to follow his propulsion after the energetic movement of which we have already spoken. The rule of the pendulum is going to be equally fulfilled here, and afterwards, a first movement (the Inferior ternary) returns to go in the opposite direction (the Superior ternary).
>
> "Humanity extends, then to reach the plane from which it has left, and for this, parting newly from the mental principle (the superior mind this time), it must ascend to the effective principle and, finally, to the spiritual.
>
> "The critical step of the evolution is, then, this mental plane which differentiates the animal from the human. While in the inferior mental, it can be classified as the concrete spirit to the man who does not think. But, in securing the better well-being, *likewise the abstract spirit* of the pure reason, it is necessary to classify it in the superior mental, with all the investigators, scientists, philosophers, etc.[4]
>
> "Until the present, only through religion has it been possible to admit to the principle of free will.

"Spiritualism offers this possibility of free will, which establishes that a man could save himself or not, the contrary of materialism. For that which is in need of importance already denies the higher-experience of individuality. Materialism was, at this point, in favor of absolute determinism, until the day in which science, representing this concrete philosophy, recognized the laws of probability.

"With the object of illustrating the question of determinism and free will, defended until today by the Materialists and the Spiritualists respectively, it is interesting to reproduce some words of the lecture delivered on December 14, 1931 by M. Louis de Lannay, President of the Academy of Sciences of Paris: The wise man who for explaining his progress, outlines the necessity of an immediate cause to every effect, would not think to consider himself as a phenomenon entirely determined to the gravitation of his former consciousness. He adds something personal which swings the balance of HIS reason. He contributes to modifying the thinking substance of the human individual, with no less efficacy than the impressions which produce in him the presence of the world or his sensations. He frees himself, with a growing liberty, to unlimited immensities, like an electric wave which vivifies all that it passes."[5]

"In what extent this inexorable determinism moves, as it comes between this liberty of action, where are rooted the foundations of these conditions, etc. . . .

"Professor Eddington considers the events of the physical world as preexistent (predetermined), and consequently, produced with restraint the lines about the universe which rules bodies. Whence determinism finds a Master erected in the physical world.

"So, then, the official teaching defending the objective orientation of materialistic science finds itself today determined to recognize the evolution of the human being, as FREE in part, from his Destiny. . . . It is necessary, however, to give here a supplementary explanation, detailed enough, since the fact of generalized relativity, as Einstein says, demands an explanation about the role of this destiny in the work of the Universe.[6]

"We can, in effect, say: though man pertains to this universe, he achieves Cosmic Ideation, in which it ought to be understood that he not only receives the influence of the third attribute (from the chain Life-Form-Thought) but, moreover, that he would have to follow a curve traced by the propulsion of his physical and psychical antecedents. . . . Much to the contrary of that set out by determinism, the human will itself can intervene in dispositions, in initiatives, in choosing the means for accomplishing his evolution.

"The fate of the ancients ought not to be understood as a mechanism, but as a law-retribution which activates the cause-effect automatism. If the flow can be focused on parts, and in other circumstances, returned to its normal streambed or, further, totally confronted, it is sufficient to have entered into activity, in its place, another principle."

"Edouard Arnaud says: 'For all thought that comes to him, man is free and responsible for the choice of this thought, and of his aspirations.'

"This, then, makes us understand, moreover, the determinism in which the mineral is sunk, for example. In fact, amidst the evolution of life, the freedom of action is made better. Thus, the savage still in the semi-bestial state, by being entirely subject to the cosmic oscillation limiting his destiny, although he comes to be in the presence of the masters, or sages, who fulfill prodigies, does not succeed in perceiving them such as they are. This is on account of the spiritual elevation of them on a much superior plane, causing them to escape all his ideation of the physical plane.

"Then it is also necessary to see the force of the law of action which has an implicit reaction in the human realm; but that in the vegetable realm it does not need that force for being plunged mathematically into the more complete determinism.

"The seat of determinism is the physical body, and its environmental medium (the conditions of space and time) is the base necessary to the fulfillment of the individuality which can be manifested by means of the inherency of the free will, whose seat is in the Essence-Life which enlivens the spiritual energy.

"In the middle of the Inspiration-Expiration Law, there is a gap which corresponds to the mental plane, and which makes the bond between the Inferior Ternary and the Superior Ternary. This state, (mental plane), indicates the setting in movement of the free will of action, which does not exist in the inferior planes of the mineral, vegetable, and animal realms. Expressed in another way, not activating this faculty of energy is recognizing its dependence on an inferior trunk . . .

"This necessity of reaction required of humanity is a normal consequence of the cosmic law. The domain of the deeds is more important than that of the laws, and for this reason, science has said:

"God orders, by the force of his mathematical Law of reestablishing the perturbed harmony. This reestablishment of the perturbed harmony by the force of the mathematical Law permits us to re-encounter the notion of mathematics, after examination through the framework of ideations and panoramas which we have seen throughout the length of this work and broadly we can comment; from mathematics we leave, and to mathematics we return.' "

The Master adds:

"And, with this, we come to the point exposed at the beginning of the prolegomena: current events are somewhat different from a simple socio-political consequence. Setting in motion the chain of energy, the propulsion comes automatically in accordance with the Involution-Evolution Law. Everything should follow the prototype rhythm because all evolution is in accord with successive cycles; nothing progresses in a continuous manner. Nevertheless, in this determinism of the universe, man is able or not to cease rolling through this cosmic fluid according to whether he is placed on a superior mental plane, or not. Still more, it is necessary to understand that the pulsations do not escape us as long as they are propulsions of activities, at frequencies of lower vibrations, or higher, than our physical senses.

"It will be noted that everything turns around the question of vibrations, and therefore, in the different worlds, according to the domain in which one is placed, we have a distinct reaction. This is valid in Life, as well as in Form, or Thought.

"To conclude, one can say that if man is not able to transform his destiny on the physical plane (and the problem will be approached even under other circumstances), as soon as he is in this body with its respiratory, digestive, and other functions, it is no less certain that he can change his psychic destiny, only that it is truly interesting, as soon as it is necessary to recognize it as the destiny of the soul.

"As a conclusion let's draw the following lesson: freedom of thought is a gift of God which permits the human being an independent activity that, naturally, involves responsibility by itself.

"Every individual has the right of using the Laws from superior planes with the object of dominating the laws of the inferior planes. This is what the truly wise men have achieved with mastery, who, by means of the exercise of polarization, are elevated to a certain degree of CAUSE with which they stop being, in the same measure, a CONSEQUENCE of the Cause as soon as having neutralized the rhythm. It is easy to escape from the operations of cause and effect in the ordinary plane.

"In the face of this freedom of action, the human being bears a great responsibility, which not only rests on his own destiny, but also includes the destiny of all humanity.

"In effect, the fault of the individual is that which he has produced in an unfortunate state of infirmity. It is equally certain that the fault of the community is that which conduces to its ruin. For all the past civilizations, the mechanism of dissolution was always the same, identical up to the point of which it could be said, as a poet has done, 'if history, which has so many books, had no more than a single page, it would come to the conclusion that, always, it is an eternal beginning . . .' Just as there are infirm individuals, there are infirm societies, and all these result from our flaws, and errors, whose weight comes to join those faults committed by our ancestors. Dr. Paul Carton expresses it thus, 'Where the human species is concerned, infirmities are declared as the accumulation of a term, and as the sanctions to collective disobediences of the natural laws of truth, goodness, unity, denial, work, sobriety, self-discipline, which are the fundamental conditions of progress, of health, and of the happiness of the human race.

"When these fundamental principles have been violated during a long time, they are stated first as a general collective infirmity. (The age of crisis is owed to imbalance of consciences.)

"This state of physical and mental degeneration is that which diminishes the vitality and morality and abates the forces of resistance of a nation. Then, the people come to a state of aptitude for contracting the infirmities classified as civil wars, revolutions, great epidemics, earthquakes, etc. . . . These scourges oblige them to suspend unwholesome habits, to find notions more just, to apply more strictly to the essential duties, (the spirit of truth, unity, discipline, charity, and altruism).

"In sum, the more the individual and his group separate from the natural law in the choice and perfection of their conditions of life and evolution, so much more dolorous are the sanctions, individual and collective, which grow because progress is a fatality. And all that which we do not want to learn well is imposed on us by the force of the right order of things. That is to say, that the partial or theoretical remedies (universal disarmament, multiple vaccinations, etc. . . .) never fully restrain the return of the scourges, because the repression of the evil, through artificial means, in a given point, would not drive it out, but make it appear in other parts, and to be transformed in greater disgrace, which would not happen if it had been restrained in its beginning.

"The obligation to discover, to proclaim, and to practice the natural laws which regulate the intellectual and moral conduct of men would remain, in the end, as the only general remedy for separating all the human miseries and, consequently, the only object or end which can be offered, as a solution to the problems of each one. In a word, good individual and social health cannot be achieved except through the good will, each time stronger, of obeying the natural and divine laws, that is to say, of thinking on works in all things with justice and goodness.

"It is perfectly logical that whoever pursues a life of materialism, having a preference for physical things, places himself in dissonance with the spirit of nature, in opposition to the soul of the collective evolution.

"The responsibility is great then, because the man who does not live united with the superior energies invading the universal harmony becomes culpable of the evils of which humanity is afflicted. On the contrary, if, through his personal rhythm, the human being elevates his thoughts to the directrix powers of the universe, he vibrates in unison with these forces, and in a constant communion with the superior rhythm of them, participates in the emission of these powerful waves, which, through his magnetism, lead the world to a superior destiny.

"We hope that these lines contribute to the birth of a better conception; that a more elevated thought is revealed to all the spirits; that there is felt the presence of the rector soul of the universe; the Sublime Thought of God."[7]

Before this paragraph of Dr. Serge Raynaud de la Ferrière, in which the concept of the divinity is united to the evolutive progression of man, his harmony, his balance, his sense of liberation accompanied by a logical argument, we are no less able to remember the harmonic correspondence of structures and the rhythm of the cosmos and man. Nor of the macrocosm and the microcosm, of the synchronous analogies of harmony like the unification of the diverse, like the concordant disposition of the discordant.

As Matila Ghyka says well, "To harmonize, or fill the interval between two given terms, consists in finding the mean which gave birth to the proportion. Plato indifferently applied these expressions to proportions of the domain of mathematics, of music, of cosmogony. The general harmonic problem consists (in the *Republic*) in putting the interval in proportion through the terms which are given in defined reasons with the initial terms, to the end of obtaining the consonance, or accord, of the intervals."[8] And in the chapter already mentioned in which Timaeus treats of the rhythm of the world soul, Plato employs the musical double tetract of the Pythagoreans, (1 + 3 + 5 + 7) + (2 + 4 + 6 + 8) = 36, the sum of the first four even numbers and the first four odd numbers, as four is for establishing the septuple celestial gamut whose tones are permitted to orchestrate the harmony of the spheres.

The influence in European thought of this passage, and of those which respectively expose the correlation between the rhythm of the world soul and the soul of man, and the morphological importance of the five regular polyhedra, was as immense in philosophy (theory of Macrocosm and of Microcosm) as in ethics and in esthetics.

Thus, we have been able to leave mathematics, and return to it, in one aspect of the primordial thought of Master Raynaud de la Ferrière and which, more than a part of his scientific thought, is a part of his universal thought.

We have not only studied with an exegetic proportion. Rather, we are animated by a social function of scientific understanding, of a table of values for a new age, that is to say, for a new humanity in which reason and wisdom guide the higher values of human transcendence.

Nevertheless, we have not approached anything but ways of following the living experience of the Master Raynaud de la Ferrière. We have not presented anything yet but the aspects of his thought which we begin to present through this series. The hundred works which he wrote are projected in time as the pristine fount which will permit the world thought interaction with his own, the perennial fount for thinkers, investigators, and writers. In other words, it is the illumined fount of a new age.

The history of fanaticisms and of limitations, of the problems of thought advanced in the face of resistance of the authority exercised without the factorial exam of circumstances and of the notions of the divinity without the criterion of an all-embracing logic of universal existence. These constitute, in part, the necessity of transposing the confused and critical period of transition between two eras. The transition of believing, which disappears through that of knowing, is raised on the horizon of our time, as it is the inexorable dawning of the mathematical movement of the spheres in the light of a new intelligence.

NOTES

1. The Great Messages. pp. 259-260.
2. ibid. p. 285.
3. ibid. p. 260.
4. ibid. Scientific Conclusion. pp. 421-422.
5. ibid. Materialism and Spiritualism. pp. 423-424.
6. Sir Arthur Eddington, The Physical World.
7. The Great Messages. Materialism and Spiritualism. pp. 424-431.
8. Matila C. Ghyka, The Golden Number. I. The Rhythms. pp. 34-5.

BIBLIOGRAPHY

ALTHUSSER LOUIS

Pour Marx. Maspero, Paris, 1965.

Para leer el capital. Siglo XXI, Madrid, 1970.

AYER, A. J.

Possibility, Probability, and Causality—Mathematics in the Behavioral Sciences.

BACHELARD, GASTON

Essai sur la connaissance approchée. Urin, Paris, 1932.

Le nouvel esprit scientifique. PUF Paris, 1934.

La dialectique de la duree. Boivin, Paris, 1936.

La psychanalyse du feu. Gallimard, Paris, 1937, 1973.

La formation de l'esprit scientifique. Urin, Paris, 1935.

La philosophie du non. PUF, 1940, 1973, Paris.

L'eau it les reves. Jose Corti, Paris, 1941.

L'air et les songes. Jose Corti, Paris, 1943.

La terre et les reveries de la volonte. Jose Corti, Paris, 1948.

La terre et les reveries du repos. Jose Corti, Paris, 1948.

Le rationalisme applique. PUF, Paris, 1948.

Le materialisme rationnel PUF, Paris 1953.

La poetique de l'espace. PUF, Paris, 1957.

La poetique de la reverie, PUF, Paris, 1960.

Lautreamont. Jose Corti, Paris, 1940.

BERGER, GASTON

Prospective Revue.

Traite practique d'analyse du caractere. PUF, Paris, 1951.

Caractere et personnalite. PUF, Paris, 1954.

BERGSON, HENRI
Essay on the immediate data of the conscience. 1925.
Matter and Memory. 1909.
Le Rire. PUF, Paris, 1900.
Creative Evolution. 1912.
L'énergie spirituelle. PUF, Paris 1919.
Les deux sources de la morale et de la religion. PUF, Paris, 1919.
La pensée et le mouvant. PUF, Paris, 1934.
Ecrits et paroles. PUF, Paris, 1957, 1958, 1959.
Oeuvres. PUF, Paris, 1963.

BERSTEIN, ALEX & ROBERTS, M. de V.
Computer versus Chess Player—Computers and Computation.
Selections from Scientific American, 1974.

BERREDO CARNEIRO, PAULO E. de
Julian Huxley, Scientist, Humanist, and Thinker. UNESCO Courier, March 1976.

BOCHENSKI JOZEF
Nove lezioni di logica simbolica. 1938.
La logique de Theophraste. 1947.

BOHR, NIELS DAVID
Wirkungs quantum und naturbeschreibung.

BRAZIER, MARY A.B.
The Analysis of Brain Waves—Mathematics in the Behavioral Sciences 1974.

CAMPELL, N.R.
Physics: The Elements, 1920.

CARNAP RUDOLF
What is Probability?—Mathematics in the Behavioral Sciences. 1974.

CARNAP, RUDOLF, O. MORGENSTERN, N. WIENER, and others Mathematics in the Behavioral Sciences.

COHEN, JOHN
Subjective Probability—Mathematics in the Behavioral Sciences. 1974.

COOPER WILLIAM W. and CHARNES, ABRAHAM
Linear Programming—Mathematics in the Behavioral Sciences. 1974.

FENICHEL, ROBERT R. and WEIZENBAUM, JOSEPH
Computers and Computation. Selections from Scientific American. 1971.

FOUCAULT, MICHEL
L'Histoire de la folie a l'age classique. Plan, Paris, 1961.
Naissance de la Clinique. PUF, Paris, 1963.
Les Mots et les choses. Gallimard, Paris, 1966.
La arqueología del saber. Siglo XXI, México, 1976.

GHYKA, MATILA C.
The Golden Number, Vol. I: The Rhythms. Buenos Aires, 1968.

GUERY, FRANCOIS
La Epistemología. La Filosofía, Ediciones Mensajero, Bilbao 1974.

HALLIDAY, DAVID & RESNICK, ROBERT
Physical Treatise II.

HEGEL, GEORG WILHELM FRIEDRICH
Fenomenología del espíritu. Tride Xizubiri, Madrid, 1935.
Science of Logic. Solar, Buenos Aires, 1969.
Encyclopedia of the Sciences of the Spirit. Tride E.O. Madrid 1918.
Propadeutic Philosophy. Minuit, Paris, 1964.
Principles of Right Philosophy. Gallimard, Paris, 1963.
Reason in History. Plon, Paris, 1965.

HEISENBERG, WERNER
The Physical Principles of the Quantum Theory. Gallimard, Paris, 1962.

HUSSERL, EDMUND
Logical Investigations. 1929.
Cartesian Meditations. J. Gaos, FCE, Mexico, 1962.
Ideas Related to Phenomenology. J. Gaos, FCE, Mexico, 1962.
The Crisis of European Humanity and Philosophy. Paylet, Paris, 1968.

HUXLEY, JULIAN
Essays of a Biologist. Knopf, New York 1929.
A Philosophy for UNESCO. The UNESCO Courier, March, 1976.
The First Years of UNESCO. The UNESCO Courier, March, 1976.

HUYGENS, CHRISTIAN
The Wave Theory of Light.

KAC, MARK
Probability—Mathematics in the Behavioral Sciences. 1974.

KANT, IMMANUEL

Critique of Pure Reason. Suares, Madrid, 1960.
Critique of Judgment. Suares, Madrid, 1958.

LEDLEY, ROBERT S. and RUDDLE, FRANK H.
The Analysis of Chromosomes by Computers. Mathematics in the Behavioral Sciences. 1974.

LOCKE, JOHN
Essays. 1974.

MACH, ERNEST
Ueber Umbeldung und anpassung in naturwissenscheftlichem denken 1883 (On Transformation and Adaptation in Scientific Cultural Thought).
Ueber das prinzip der Vergleichung in der physik. 1894 (On the Principle of Comparison in Physics).
Erkenntnis und irrtum. Skizzen sur Psychologis, der forschung. 1905. (Skill and Error. Searches for the Psychology of Research).
Die leitgedanken meiner naturwissenscheftlichem erkenniskehere und ihre aufnahme durch die zeitgenossen. 1919 (The Directing Ideas of my Theory of Scientific-Natural Thought and the Reactions of my Contemporaries in the Face of Them).

McCARTHY, JOHN
Information—Computers and Computation. Selections from Scientific American, 1971.

MEYERSON, EMILE
Identity and Reality. 1929.

PERROUX, FRANCOIS
Perroux Interrogates Marcuse. 1969.

PIAGET, JEAN
Traite de logique. A. Colin, Paris, 1949.
Introduction a l'epistemologie genetique. PUF, Paris, 1950.
Sagesse et illusions de la philosophie. PUF, Paris, 1965.

RAYNAUD de la FERRIÈRE, SERGE
Great Messages. Series of 7 Tomes. Editorial Diana Mexico, 1976.
The Mysteries Revealed. Editorial Diana, Mexico.
Mysticism in the 20th Century. Editorial Diana, Mexico.
Art in the New Age. Editorial Diana, Mexico.
Science and Religion. Editorial Diana, Mexico.
Black Book of Freemasonry. Editorial Diana, Mexico.
A Mathesis of Psychology. Editorial Diana, Mexico.

His Circulars and Letters from France, Tomes I, II, III. UGB Perú.

Psychological Talks. Series of 36 Tomes. UGB, Perú.

SAINT IVES d'ALVEYDRE (Marquis)

L'Archeometra. Dorbon Aine, Paris, 1905.

SERRES, MICHEL

Les anamneses mathematiques. 1967.

SIMONDON, GILBERT

Du mode d'existence des objects techniques. Aubier Montaigne, Paris, 1958.

SUPPES, PATRICK

The Use of Computers in Education: Mathématics in the Behavioral Sciences. 1974.

SUTHERLAND, IVAN E.

Entrances and Exits.

Systems of Visual Representation: Computers and Computation. 1974.

TEJADA RODRIGUEZ, ROBERTO

The Magic Squares and Their Projection on the 20th Century. Peru.

VON NEUMANN—ROSS, ASHBY W.

Introduction to Cybernetics. Buenos Aires, 1960.

Project for a Brain. Tecnos, Madrid, 1965.

WHITEHEAD, ALFRED NORTH

Science and the Modern World. 1926.

Principia Mathematica, I. 1910; II. 1912.

WIENER, NORBERT

Cybernetics: Mathematics in the Behavioral Sciences. 1974.

Cybernetics or Control and Communication in the Animal and the Machine. 1948.

WILLIAMS, H.S.

The Elements of the Geological. Times Scales J. Geol. 1893.